# Magnetic Sensors and Applications Based on Thin Magnetically Soft Wires with Tunable Magnetic Properties

A. Zhukov and V. Zhukova

# Magnetic Sensors and Applications Based on Thin Magnetically Soft Wires with Tunable Magnetic Properties

International Frequency Sensor Association Publishing

*A. Zhukov and V. Zhukova*
Magnetic Sensors and Applications Based on Thin Magnetically Soft Wires
with Tunable Magnetic Properties

ISBN-10: 84-617-1087-8
ISBN-13: 978-84-617-1087-4
BN-20140716-XX
BIC: TJFD

# About the Authors

**Prof. Arcady P. Zhukov** is a research professor of the IKERBASQUE, Basque Foundation for Science (Spain). He was graduated from the Physics-Chemistry Department of Moscow Steel and Alloys Institute in 1980. Since 1980 he was working in Institute of Solid State Physics of Russian Academy of Sciences (1980-1994), the Instituto de Magnetismo Aplicado (Madrid, Spain) in 1994-1997, the University of Basque Country in 1997-2001 and 2003-2010, in the Donostia International Physics Centre (2001-2002), in the Institute of Material Science, CSIC (2002- 2003). He obtained the Ph.D. in 1988, habilitation (Doctor of Science degree) -in 2010. He published about 400 referred papers in the international ISI journal on magnetic materials, edited a conference proceedings, gave a number of invited talks on few international conferences on Magnetism, wrote a book (together with V. Zhukova) "Magnetic properties and applications of ferromagnetic microwires with amorphous and nanocrystalline structure", few books chapters in the book "Advanced Magnetic Materials", for the "Encyclopedia of NanoScience and Nanotechnology", "Encyclopedia of Sensors". He organized few International conferences, among them Joint International Magnetic Symposium", JEMS′06 (San Sebastián, June 2006, co-chair) and Donostia International Conference on Nanoscaled Magnetism and Applications (DICNMA 2013, co-chair).

**Dr. Valentina Zhukova** is a researcher of the Dept. of Material Physics, Basque Country University, Spain. Graduated as an engineer in the Metallurgy in Moscow Steel and Alloys Institute (presently National University of Science and Technology) in 1982 and received PhD degree in 2003 in the Basque Country University, Spain in the studies of transport and magnetic properties of glass-coating microwires. Current fields of interest: amorphous and nanocrystalline ferromagnetic materials, magnetic micro-wires, giant magneto-impedance, giant magnetoresistance, magnetoelastic sensors. She has published more than 180 referred papers in the international journals, few book chapters, author of two patents, and participates in research projects. She was a Co-Chairman of the "International Workshop on Magnetic Wires – 2008" (Zumaia, Mayo 2008), member of the Organizing Committees in: "International Workshop on Magnetic

Wires – 2001" (San Sebastián, June 2001), "Joint European Magnetic symposium" JEMS-2006 (San Sebastián, June 2006) and Donostia International Conference on Nanoscaled Magnetism and Applications (DICNMA 2013, Programme Committee).

# Contents

# Preface

This book on magnetic microwires for magnetic sensors applications is inspired by a rapidly growing interest in the development of functional materials with improved magnetic and magneto-transport properties and in sensitive and inexpensive magnetic sensors. The research is demanded by the last advances in technology and engineering. Certain industrial sectors, such as magnetic sensors, microelectronics or security demand cost-effective materials with reduced dimensionality and desirable magnetic properties (i.e., enhanced magnetic softness, giant magnetic field sensitivity, fast magnetization switching etc.). Consequently, the development of soft magnetic materials in different forms of ribbons, wires, microwires, and multilayered thin films continue to attract significant attention from the scientific community, as the discovery of the so-called giant magnetoimpedance effect in these materials makes them very attractive for a wide range of high-performance sensor applications ranging from engineering, industry to biomedicine.

One of the recent tendencies related with development of industrial applications in the field of magnetic sensors is the miniaturization of the magnetic sensors. This tendency stimulated development of technology for magnetic materials with reduced dimensionality, such as thin films and thin wires. Consequently, since 2010 glass-coated metallic microwires exhibiting giant magnetoimpedance effect are using in real technological applications for low magnetic field detection owing to high magnetic field sensitivity.

This book aims to provide most up-to-date information about recent developments in magnetic microwires for advanced technologies and present recent results on the remagnetization process, domain walls dynamics, compositional dependence and processing of glass-coated microwires with amorphous and nanocrystalline character suitable for magnetic sensors applications.

We hope this book will stimulate further interest in magnetic materials research and that this book can be of interest for PhD students, postdoctoral students and researchers working in the field of soft magnetic materials and applications.

Last but not least, we would like to acknowledge our colleagues and collaborators for great contributions and assistance to this book preparation.

# 1. Introduction

Recent technological advances are greatly affected by the development of advanced functional materials with improved physical properties. Advanced magnetic materials form important part among the functional materials. Many industrial sectors, such as magnetic sensors, microelectronics, security, automobile, energy-efficient refrigerators, medicine, aerospace, energy harvesting and conversion, informatics, electrical engineering, magnetic recording etc demand cost-effective magnetic materials.

It is worth mentioning that the topic of Magnetic Materials is of a highly interdisciplinary nature and combines features of crystal chemistry, metallurgy, and solid state physics.

Magnetic materials are broadly classified into two main groups with either hard or soft magnetic characteristics. Soft magnetic materials can be magnetized by relatively low-strength magnetic fields, and when the applied field is removed, they return to a state of relatively low residual magnetism. Soft magnetic materials typically exhibit coercivities values, Hc, of approximately 400 A/m (5 Oe) to as low as 0.16 A/ m (0.002 Oe). Soft magnetic behavior is important in any application involving a change in magnetic induction. Hard magnetic materials retain a large amount of residual magnetism after exposure to a magnetic field. These materials typically have coercivities, Hc, of 10 kA/m (125 Oe) to 1 MA/m (12 kOe). These materials are used principally as a source of a magnetic field.

Soft-magnetic materials are characterized by low coercivities, high values of magnetic permeability (initial permeability $\mu_a \sim 102 - 105$; maximum permeability $\mu_{max} \sim 103\text{-}106$), and low magnetic hysteresis losses per remagnetization cycle are very small. These properties of soft-magnetic materials are related with the low magnetocrystalline anisotropy (such as Fe-Ni-based soft-magnetic materials, amorphous magnetic materials and some ferrites), also from a low magnitude of the magnetostriction constant, $\lambda_s$.

Soft magnetic materials are used extensively in power electronic circuits, as voltage and current transformers, saturable reactors, magnetic amplifiers, inductors, chokes and magnetic sensors.

One of most promising families soft magnetic materials are amorphous magnetic materials introduced few decades ago [1, 2]. The main interest in amorphous soft magnetic materials is related with their liquid-like structure characterized by the absence of long range ordering. Particularly the absence of magnetocrystalline anisotropy is the main reason of extremely soft magnetic properties exhibited by amorphous magnetic materials [2-4].

First amorphous materials have been produced more than 50 years ago by rapid quenching from the liquid state by I. S. Miroshnitchenko and I. V. Salli [5] and later by P. Duwez et al [6]. Technological development of the fabrication technique and studies of the structure, glass formation ability and thermodynamics and magnetism of amorphous alloys were intensively performed in 60-th-70-th.

They are presently used in many areas, including transformers for electric power distribution, power electronics for small and large-scale power management, pulse power devices, telecommunication devices, and sensors [3, 4]. The energy efficient nature of amorphous magnetic materials has a great impact on global energy savings when large devices such as transformers are in use. The ubiquitous use of power electronics in information technologies demand ever-more efficient electronic devices in which amorphous soft magnets play a major role. These devices, however, introduce harmonic distortions in the electrical power lines, which in turn increase total transformer losses. Although this has been recognized, its impact on the total losses of conventional transformers has been found to be considerably larger than on amorphous metal-based transformers [1]. Thus, the use of amorphous metal-based electrical transformers is becoming increasingly significant. In the power electronics area, small-size saturable cores using non-magnetostrictive amorphous alloys have been widely utilized as magnetic amplifiers to control output voltages of switch-mode power supplies in PCs. The trend in the small-size power supplies is toward dc-to-dc conversion with low output voltages (~1 volt) and high currents (10-100 A), which requires low-loss electrical chokes. These components are usually based on high induction Fe-based alloys [2]. The magnetic characteristics of these components can be utilized in power factor correction to maximize electrical power usage in power conditioning devices.

Therefore, the development of soft magnetic materials in different forms of ribbons, wires, microwires, and multilayered thin films with amorphous and nanocrystalline structure continue to attract significant attention of the scientific community, as the discovery of the so-called giant magnetoimpedance effect in these materials makes them very attractive for a wide range of high-performance sensor applications in the field of engineering, industry or biomedicine [2, 5-8]. In another research area, the development of advanced magnetocaloric materials for magnetic refrigeration technology has also generated growing interest among magnetic community. The majority of magnetic refrigeration is to develop new materials that are cost-effective and possess high cooling efficiencies (i.e. large magnetocaloric effect over a wide temperature range). In all cases, a comprehensive understanding of the processing-structure-property relationship in the fabricated materials is of critical importance.

One of the recent tendencies related with development of industrial applications in the field of magnetic sensors is the miniaturization of the magnetic sensors. Certain progress has been recently achieved in fabrication of novel magnetic nano-materials (thin films, nanowires, nano-dots...), but at the same time quite sophisticated technology should be used but in many occasions the magnetic properties of these materials are rather poorer than such properties of bulk magnetic materials (amorphous ribbons, wires, sintered materials...) and the fabrication process is much more expansive and complex [1, 2, 8]. On the other hand certain industrial sectors, like magnetic sensors, microelectronics, security etc, need cheap materials with reduced dimensionality and simultaneously with high magnetic properties (particularly enhanced magnetic softness). Therefore reduction of the dimensionality of magnetically soft materials is one of the priority tasks in the field of applied magnetism. This tendency stimulated development of technology for magnetic materials with reduced dimensionality, such as thin films and thin wires [2, 5-11]. Consequently, soft magnetic wires with reduced dimensionality and outstanding magnetic characteristics, such as melt extracted wires (typically with diameters of 40-50 μm) [9] and glass-coated microwires with even thinner diameters (between 1-40 μm) [5-9] recently gained much attention. The advantage of the Taylor-Ulitovsky method allowing the fabrication of glass-coated metallic microwires consists of controllable fabrication of long (up to few km long continuous microwires), homogeneous, rather thin and composite wires.

Although initially reported magnetic softness was clearly poorer than of thicker amorphous alloys (ribbons and wires), recently very soft magnetic properties have been reported [8, 9]. This gives rise to development of industrial applications for low magnetic field detection in various industrial sectors [5-9, 12, 13].

GMI effect, consisting of large sensitivity of the impedance of magnetically soft conductor on applied magnetic field, has been successfully explained in the terms of classical electrodynamics through the influence of magnetic field on penetration depth of electrical current flowing through the magnetically soft conductor attracted great attention in the field of applied magnetism [14-16] basically due to excellent magnetic field sensitivity suitable for low magnetic field detection.

Cylindrical shape and high circumferential permeability observed in amorphous wires are quite favourable for achievement of high GMI effect [9-12, 14-16]. As a rule, better soft magnetic properties are observed for nearly-zero magnetostrictive compositions. It is worth mentioning, that the magnetostriction constant, $\lambda_s$, in system $(Co_xFe_{1-x})_{75}Si_{15}B_{10}$ changes with x from $-5\times10^{-6}$ at x = 1, to $\lambda_s\approx35 \times10^{-6}$ at x$\approx$0.2, achieving nearly-zero values at Co/Fe about 70/5 [17-19].

It was pointed out [7-12, 20] that the good magnetic softness is directly related to the GMI effect: the magnetic field dependence of the GMI spectra is mainly determined by the type of magnetic anisotropy. Thus the circumferential anisotropy leads to the observation of the maximum of the real component of wire impedance (and consequently of the GMI ratio) as a function of the external magnetic field. On the other hand, in the case of axial magnetic anisotropy the maximum value of the GMI ratio corresponds to zero magnetic fields [7-9, 20], i.e. results in a monotonic decay of the GMI ratio with the axial magnetic field. Like magnetic permeability, GMI effect presents tensor character. Consequently the "scalar" model of GMI effect was significantly modified taking into account the tensor origin of the magnetic permeability and magneto impedance [20, 21]. Off-diagonal components of the magnetic permeability tensor and impedance tensor were introduced [20-22] in order to describe the GMI effect in amorphous wires with circumferential magnetic anisotropy. It was established that to achieve high GMI effect, the magnetic anisotropy should be as small as possible.

From the point of view of industrial applications low hysteretic and linear magnetic field dependence of the output signal are desirable [10-13, 23]. Anti-symmetrical magnetic field dependence of the output voltage with linear region has been obtained for pulsed GMI effect based on detection of the off-diagonal GMI component of amorphous wires [8, 10, 21]. Such pulsed scheme for GMI measurements resulted quite useful for real GMI sensors development [12, 13].

On the other hand microwires with Fe-rich metallic nucleus composition present completely different magnetic properties exhibiting rectangular hysteresis loops related with large and single Barkhausen jump. The remagnetization of this family of microwires is determined by the fast magnetization switching within single domain axially magnetized inner core [9, 18]. It is worth mentioning, that recently controllable and fast domain wall (DW) propagation observed in various families of thin magnetic wires prepared by different methods [24, 25] has been proposed for the information storage, magnetic sensors and logics [24-26]. Certainly the DW speed and the possibilities for the DW dynamics manipulation are quite relevant for these applications.

For the magnetic field driven DW propagation the magnetic field value and the wire dimensions affect the DW velocity [24-27]. Rather fast DW velocity usually exceeding 1000 m/s has been reported for amorphous micrometric wires with cylindrical cross section and positive magnetostriction constant [26, 28]. On the other hand, several methods for controlling the DWs dynamics in nanowires (usually with rectangular cross-section), such as DW injection, creation of artificial defects and inhomogeneities have been reported [24, 25].

As mentioned above, in the case of magnetic microwires with positive magnetostriction constant the magnetic bistability characterized by the appearance of rectangular hysteresis loop at low applied magnetic field has been observed [8, 18, 29]. This magnetic bistable behavior is related to the presence of a single Large Barkhausen Jump, which was interpreted as the magnetization reversal in a large single domain [8]. It is important, that single and large Barkhausen jump is observed above some critical fields. As regarding the critical length, it can be correlated well with the demagnetizing factor [29] indicating that the closure domains penetrate from the wire ends inside the internal axially magnetized core destroying the single domain structure. For the case of commercially available amorphous wires (with diameter about 120 μm) this critical length is about 7 cm, which is quite inconvenient for use in magnetic micro-sensors and microelectronics. In glass-coated microwires with diameter about 10 μm this critical length is much shorter (about 2 mm) which is quite suitable for applications in microsensors. This rectangular hysteresis loop also disappears when the magnetic field is below some critical value denominated as the switching field [29]. Such rectangular hysteresis loop was interpreted in terms of nucleation or depinning of the reversed domains inside the internal single domain and the consequent domain wall propagation [26, 29]. Perfectly rectangular shape of the hysteresis loop has been related with a very high velocity of such domain wall propagation. It is demonstrated by few methods that the remagnetization process of such magnetic microwire starts from the sample ends as a consequence of the depinning of the domain walls and subsequent DW propagation from the closure domains [26, 29].

Consequently amorphous glass-coated microwires with positive magnetostriction constant are unique material allowing studying the magnetization dynamics of a single DW in a cylindrical micrometric wire. This peculiar magnetization switching is related with their domain structure determined by the stress distribution during rapid solidification fabrication process [5-9]. The magnetization switching is therefore related with the propagation of the single head-to head DW along the wire [26, 28].

On the other hand development of a new types of stress- and temperature-tunable meta- materials consisted of short pieces of conductive ferromagnetic wires embedded into a dielectric matrix with the effective microwave permittivity depending on an external dc magnetic field, applied stress or temperature recently have been reported [30-33]. The short wire inclusions play a role of "the elementary scatterers", when the electromagnetic wave irradiates the composite and induces a longitudinal current distribution and electrical dipole moment in each inclusion. These induced dipole moments form the dipole response, which can be characterized by some complex effective permittivity. The later may have a resonance or relaxation dispersion caused by the strong current distribution along a wire, which depends on the wire high frequency surface impedance. For a ferromagnetic conductive wire, the surface impedance may depend not only on its conductivity but also on the dc external magnetic field and tension through the GMI effect. Therefore, the dispersion of the effective permittivity can be tuned from a resonance type to a relaxation type, when a sufficient magnetic field or tensile stress is

16

applied to the composite sample. It is worth mentioning that thin wires with stress sensitive magnetic anisotropy exhibiting stress sensitive GMI effect and SI effect are quite necessary for designing of such composites.

Consequently studies of thin magnetic wires with reduced geometrical dimensions (of order of 1-30 $\mu$m in diameter) gained importance within last few years [5-9]. Several exciting results on excellent soft magnetic properties (with coercivities till 4 A/m), extremely high and low hysteretic Giant Magneto-impedance effect, GMI, and fast DW propagation in micrometric amorphous and nanocrystalline wires have been reported [5-9, 22, 23].

The main application of thin magnetically soft wires is based on extremely high GMI effect. On the other hand, existing and future applications of such thin wires do not restricted by the GMI-based applications. Thus, a number of magnetic sensors based on giant magneto-impedance (GMI) effect and stress-impedance (SI) effect with the C-MOS IC circuitry and advantageous features comparing with conventional magnetic sensors have been reported [7-9]. Main proposed applications are related with the detection of the magnetic fields, small weights and vibrations, and such branches of the industry as the car industry and medicine are main consumers of these sensors [7-9].

Although crystallization of amorphous materials usually results in degradation of their magnetic softness, in some cases crystallization can improve magnetically soft behaviour. This is the case of so-called "nanocrystalline" alloys obtained by suitable annealing of amorphous metals. These materials have been introduced in 1988 by Yoshizawa et al. [34] and later have been intensively studied by a number of research groups [35-37]. Research and technological interest in such nanocrystalline alloys, denominated also as "Finemet" (in the case of Fe-rich nanocrystalline alloys) arose from extremely soft magnetic properties combined with high saturation magnetization. This nanocrystalline structure of partially crystalline amorphous precursor is observed in Fe-Si-B with small additions of Cu and Nb. It is widely assumed that the role of these small additions of Cu and Nb results in inhibiting of the grains nucleation and decreasing of the grain growth rate [34-36]. Such soft magnetic character is thought to be originated because the magnetocrystalline anisotropy vanishes and the very small magnetostriction value when the grain size approaches 10 nm [34-36]. As was theoretically estimated by Herzer [35], average anisotropy for randomly oriented $\alpha$-Fe (Si) grains is negligibly small when grain diameter does not exceed about 10 nm. In addition to the suppressed magnetocrystalline anisotropy, low magnetostriction values provide the basis for the superior soft magnetic properties observed in particular compositions. Low values of the saturation magnetostriction are essential to avoid magnetoelastic anisotropies arising from internal or external mechanical stresses [37].

As mentioned above, amorphous wires have been studied starting from 90-th [38-39]. First generation of amorphous wire deals with typical diameter around 125 µm in diameter, obtained by the so-called in-rotating-water quenching technique. This kind of materials exhibits a number of unusual magnetic properties. Thus, the magnetostrictive compositions exhibit rectangular hysteresis loop, while best magnetic softness is observed for the nearly-zero magnetostriction composition. Their main technological interest is related to the magnetic softness in nearly-zero magnetostriction composition, magnetic bistability in non-zero magnetostriction compositions and aforementioned GMI effect [14, 15, 38, 39].

The alternative technology of rapid quenching – Taylor-Ulitovsky method to produce thinner metallic wires (in the order of 1 to 30 µm in diameter) covered by an insulating glass coating is known along many years [40-43], but has been widely employed for fabrication of ferromagnetic amorphous microwires coated by glass (see photo in Fig. 1) since middle of 90-th [5-9]. The fabrication method denominated in most of modern publications as a modified Taylor-Ulitovsky and/or quenching-and-drawing method is actually well-known since 60-th and well described in Russian in 60-th [40,42] as well as in recent publications [18].

In the laboratory process, an ingot containing few grams of the master alloy with the desired composition is placed into a Pyrex-like glass tube and within a high frequency inductor heater. The alloy is heated up to its melting point, forming a droplet. While the metal melts, the portion of the glass tube adjacent to the melting metal softens, enveloping the metal droplet. A glass capillary is then drawn from the softened glass portion and wound on a rotating coil. At suitable drawing conditions, the molten metal fills the glass capillary and a microwire is thus formed where the metal core is completely coated by a glass shell.

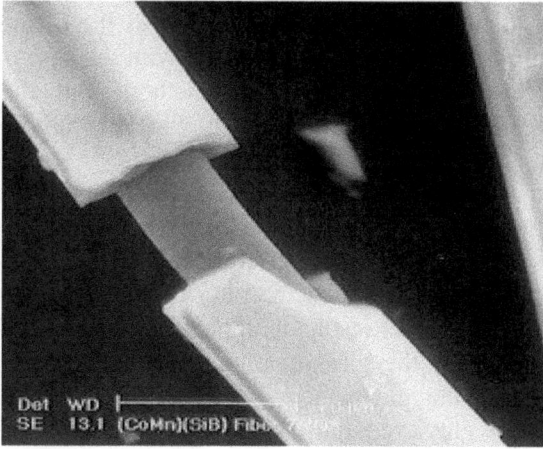

**Figure 1.** Micrograph of the glass-coated microwire. Reprinted with permission from [8] V. Zhukova, M. Ipatov and A Zhukov, Thin Magnetically Soft Wires for Magnetic Microsensors, *Sensors*, 9, 9216-9240, 2009, doi:10.3390/s91109216 (Fig. 2).

The microstructure of a microwire (and hence, its properties) depends mainly on the cooling rate. Chemical and metallurgical processes related with interaction of the ingot alloy and the glass, electromagnetic and electro-hydrodynamic phenomena in the system of inductor- ingot, thermal conditions of formation of cast microwire, parameters of the casting process and their limits affecting the casting rate and the diameter of a microwire were describe in details for non-magnetic microwires [40-42] and overviewed in relatively recent book [18]. From the point of view of magnetic properties of thin magnetic microwires and properties related with surface layers (like is the case of GMI effect) the interfacial layer between the metallic nucleus and glass coating is especially relevant [18]. The features of the interfacial layer between the metallic nucleus and glass coating (it thickness, structure and physical properties) depend on the origin of the interfacial layer. Thus, the thickness of the interfacial layer might have from few µm for the case of the formation of the series of solid solutions or stable chemical compounds to less than 0.1 µm in the case of the origin related with the uncompensated molecular forces on the interface between the glass and the metallic nucleus [18, 40-42].

The other source of instability of properties of cast microwire is related with gas content inside the microwire. The sources of the gas are: the atmosphere, the gas impurities in the alloy and the glass.

The great advantage of these microwires is that the obtained diameter could be significantly reduced as-compared with the case of amorphous wires produced by in-rotating water method. But their magnetic properties are also quite different from "thicker "amorphous wires. Thus although like in the case of "thicker wires" it was observed that Fe-rich compositions with positive magnetostriction constant show generally rectangular hysteresis loop, Co-rich negative magnetostrictive compositions present completely different character of hysteresis loops. Co-rich microwires have almost non-hysteretic magnetization curves and glass coating removal results in appearance of magnetic bistability [18]. This is because the glass coating introduces additional internal stresses due to the difference between the thermal expansion coefficients of glass coating and metallic nucleus. Therefore, microwires of the same composition can show different magnetic properties because of the different magnetoelastic energy. Depending on the thickness of the glass coating, the switching field (applied magnetic field necessary to observe magnetic bistability) is generally one order of magnitude higher than for melt-spun wires.

On the other hand, when the magnetostriction constant, $\lambda_s$, is close to zero, a great variety of magnetic effects can be observed, depending on the sign of $\lambda_s$. It has been found that the hysteresis loop changes from unhysteretic for slightly negative magnetostriction constant to rectangular for positive magnetostriction [44].

The internal stresses results to be as the main source of magnetic anisotropy in amorphous and nanocrystalline materials due to the magnetoelastic coupling between magnetization and internal stresses through magnetostriction and absence of crystalline structure and defects typical for crystalline materials (grain boundaries, dislocations...). In most of amorphous materials the main origin of these internal stresses is related with high quenching rate and solidification process. Thus solidification usually proceeds from the surface. In the case of amorphous ribbons additionally the cooling rate is higher from the contact side of ribbon, i.e. where the ribbons contacts with the quenching drum [2, 23, 27]. As a result, the morphology of contact and free surfaces of ribbons are quite different too.

In the case of glass-coated microwires, the fabrication process is different. It involves simultaneous solidification of metallic nucleus surrounded by the glass coating. This introduces an additional magnetoelastic contribution to the magnetic anisotropy acting as a new parameter determining the magnetization process [8, 9]. The origin of these additional stresses is determined by significant difference of the thermal expansion coefficients of the glass and the metal [7, 8, 18, 45-47].

Recent studies demonstrated that the same fabrication technique for fabrication of glass-coated thin wires allows obtaining of microwires with granular structure exhibiting giant magnetoresistance (GMR) effect [48, 49], Heusler-type microwires [50, 51] microwires with magnetocaloric effect [52, 53]. In latter cases the composition of metallic nucleus was different from the case of amorphous magnetically soft microwires. Consequently studies of thin magnetic wires gain more and more attention.

In this book we are paying attention on overview of fabrication, processing and tailoring of magnetic properties of amorphous microwires exhibiting a number of exciting functional magnetic properties interesting for already introduced and proposed applications in magnetic sensors.

# 2. Giant Magneto-impedance Effect

As already mentioned in the introduction, the GMI effect usually observed in soft magnetic materials phenomenologically consists of the change of the AC impedance, $Z = R + iX$ (where R is the real part, or resistance, and $X$ is the imaginary part, or reactance), when submitted to an external magnetic field, $H_0$. The GMI effect was well interpreted in terms of the classical skin effect in a magnetic conductor assuming the dependence of the penetration depth of the *ac* current flowing through the magnetically soft conductor on the *dc* applied magnetic field [14-16]. Extremely high sensitivity of the GMI effect to even low magnetic field attracted great interest in the field of applied magnetism basically for applications for low magnetic field detection.

Generally, the GMI effect was interpreted assuming scalar character for the magnetic permeability, as a consequence of the change in the penetration depth of the *ac* current caused by the *dc* applied magnetic field. The electrical impedance, $Z$, of a magnetic conductor in this case is given by [14-16]:

$$Z = R_{dc}\, krJ_0(kr)\big/2J_1(kr) \qquad (1)$$

with $k = (1 + j)/\delta$, where $J_0$ and $J_1$ are the Bessel functions, $r$ is the wire's radius and $\delta$ the penetration depth given by:

$$\delta = 1\big/\sqrt{\pi\sigma\mu_\phi f} \qquad (2)$$

where $\sigma$ is the electrical conductivity, $f$ is the frequency of the current along the sample, and $\mu\phi$ is the circular magnetic permeability assumed to be scalar. The *dc* applied magnetic field introduces significant changes in the circular permeability, $\mu\phi$. Therefore, the penetration depth also changes through and finally results in a change of $Z$ [14-16].

Usually for quantification of the GMI effect the magneto impedance ratio, $\Delta Z/Z$, is used. GMI ratio, $\Delta Z/Z$, is defined as:

$$\Delta Z/Z = [Z\,(H) - Z\,(H_{max})]\,/\,Z\,(H_{max}), \qquad (3)$$

where $H_{max}$ is the axial *DC*-field with maximum value up to few kA/m.

The main features of the GMI effect are the following:

1. Large change in the total impedance usually above 100 %. Usually for the case of amorphous wires with high circumferential permeability the highest GMI effect is reported [18]. Thus, few researchers reported on achievement of about 600 % GMI ratio in Co-rich microwires with vanishing magnetostriction constant [54, 55]. In this case, it is quite promising for the application of magnetic sensors.

2. The GMI materials, whether wires, ribbons or films, are usually extremely soft magnetic materials. It was pointed out [2-6] that the good magnetic softness is directly related to the GMI effect: the magnetic field dependence of the GMI spectra is mainly determined by the type of magnetic anisotropy. Thus the circumferential anisotropy leads to the observation of the maximum of the real component of wire impedance (and consequently of the GMI ratio) as a function of the external magnetic field. On the other hand, in the case of axial magnetic anisotropy the maximum value of the GMI ratio corresponds to zero magnetic fields [2-6], i.e. results in a monotonic decay of the GMI ratio with the axial magnetic field.

3. The alternating current plays an important part in the GMI effect. The main reason is that like magnetic permeability, GMI effect presents tensor character [56, 57]. Therefore AC current flowing through the sample creates circumferential magnetic field. Additionally AC current produces the Joule heating [58]. There are many publications related with the origin of the GMI effect [12-16, 59-61]. It must be underlined, that the GMI effect origin has been explained based on the theory of classical electrodynamics. The skin effect, which is responsible for GMI at medium and high frequencies, is a phenomenon well described by the classical electrodynamics [62] many years ago. As a consequence of induced eddy currents, the high frequency AC current is not uniformly distributed in the conductor volume but is confined to a shell close to the surface, with depth, $\delta$, given by eq. (2).

Cylindrical shape and high circumferential permeability observed in amorphous wires are quite favorable for achievement of high GMI effect [20, 21]. As a rule, better soft magnetic properties are observed for nearly-zero magnetostrictive compositions. It is worth mentioning, that the magnetostriction constant, $\lambda s$, in system $(Co_xFe_{1-x})_{75}Si_{15}B_{10}$ changes with x from $-5 \times 10^{-6}$ at x= 1, to $\lambda s \approx 35 \times 10^{-6}$ at x$\approx$0.2, achieving nearly-zero values at Co/Fe about 70/5 [17, 19, 63].

Depending on the frequency $f$ of the driving AC current $Iac$ flowing through the sample, the giant magnetoimpedance can be roughly four different regimes might be considered. In fact we should consider mostly comparison of the skin depth with the radius or half thickness of the sample:

(*i*) At low frequency range of 1-10 kHz when the skin depth is larger than the radius or half thickness of the sample (rather weak skin effect) the Matteucci effect and magnetoinductive effect have been observed. In particular sharp voltage peaks measured between the sample's ends were attributed to the sample remagnetization considering circular magnetization reversal [65-69]. The changes of impedance are due to a circular magnetization process exclusively. Therefore considering that the origin of GMI effect is associated with the skin effect of magnetic conductor, observed phenomena might not be considered properly as the GMI effect.

(*ii*) At frequencies, ranging from 10-100 kHz to 1-10 MHz, the frequency range where the GMI effect has been firstly reported and described, the giant magnetoimpedance, originates basically from variations of the magnetic penetration depth due to strong changes of the effective magnetic permeability caused by a DC magnetic field [14, 15]. It is widely believed that in this case both domain walls and magnetization rotation contribute to changes of the circular permeability and consequently to the skin effect.

(*iii*) For frequencies ranging in the MHz band (from 1-10 MHz to 100-1000 MHz depending on the geometry of the sample), the skin effect is also the originated by the skin effect of the soft magnetic conductor, i.e. must be attributed to the GMI. But at these frequencies the domain walls are strongly damped. Therefore the magnetization rotation must be considered as responsible for the magnetic permeability change induced by an external magnetic field [70].

(*iv*) At high frequencies, of the order of GHz, the magnetization rotation is strongly influenced by the gyromagnetic effect. With increasing the frequency the GMI peaks are shifted to static fields where sample is magnetically saturated. At his frequency range strong changes of the sample's impedance have been attributed to the ferromagnetic resonance (FMR) [71, 72].

We must underline that the criterion used for determining the frequency regions is somehow artificial and, to some extent it is rather rough and arbitrary. Thus, the most appropriate criteria is probably the ratio of skin depth to transversal dimensions of the sample ($d/a$), used by most authors [73]. In this case, the criteria used must be the ratio $d/a$:

$d/a \gg 1$ indicate a weak skin effect regime, while $d/a \ll 1$ indicate a strong skin effect.

Generally, weak skin effect is observed at lower frequencies. However, $d/a$ ratio depend on many other parameters, such as sample dimensions, material properties, magnetic field, etc. Therefore, this criterion also does not seem to be appropriate for the distinction among different frequency ranges.

The other important parameter is the domain walls contribution: are they damped or not.

Although different regimes can be roughly separated, the physical origin is essentially the same [59]. In fact, the skin effect is also responsible for FMR absorption in ferromagnetic metals [74].

In ferromagnetic materials with high circumferential anisotropy (the case of magnetic wires) the magnetic permeability possesses the tensor nature and the classic form of impedance definition is no valid. The relation between electric field (which determines the voltage) and the magnetic field (which determines the current) is defined through the surface impedance tensor [61, 75]

$$e = \hat{\varsigma} h \quad \text{or} \quad \left\{ \begin{aligned} e_z &= \varsigma_{zz} h_\varphi - \varsigma_{z\varphi} h_z \\ e_\varphi &= \varsigma_{\varphi z} h_\varphi - \varsigma_{\varphi\varphi} h_z \end{aligned} \right\} \quad (4)$$

The circular magnetic fields $h_\varphi$ is produced by the currents $i_w$ running through the wire. At the wire surface $h_z = i/2\pi r$, where $r$ is the wire radius. The longitudinal magnetic fields $h_z$ is produced by the currents $i_c$ running through the exciting coil, $h_z = N_l i_c$, where $N_l$ is the exciting coil number of turns.

Consequently the "scalar" model of GMI effect was significantly modified taking into account the tensor origin of the magnetic permeability and magneto impedance [8, 21-23, 56, 57, 75-78]. Non-diagonal components of the magnetic permeability tensor and impedance tensor were introduced [20-23, 56, 57, 61, 75-78] in order to describe the circumferential magnetic anisotropy in amorphous wires.

It was established that to achieve high GMI effect, the magnetic anisotropy should be as small as possible [20].

Various excitation and measurement methods are required to reveal the impedance matrix elements. The longitudinal and circumferential electrical field on the wire surface can be measured as voltage drop along the wire $v_w$ and voltage induced in the pickup coil $v_c$ wound on it [21-23, 75-78].

$$v_w \equiv e_z l_w = (\varsigma_{zz} h_\varphi - \varsigma_{z\varphi} h_z) l_w \qquad (5)$$

$$v_c \equiv e_\varphi l_t = (\varsigma_{\varphi z} h_\varphi - \varsigma_{\varphi\varphi} h_z) l_t, \qquad (6)$$

where $l_w$ is the wire length, $l_t = 2\pi r N_2$ is the total length of the pickup coil turns $N_2$ wounded directly on the wire.

The methods for revealing the different elements of impedance tensor are shown in Fig. 2. The longitudinal diagonal component $\varsigma_{zz}$ defines as the voltage drop along the wire and corresponds to impedance definition in classical model (Fig. 2.a)

$$\varsigma_{zz} \equiv \frac{v_w}{h_\varphi l_w} = \left(\frac{2\pi a}{l_w}\right)\left(\frac{v_w}{i_w}\right) \qquad (7)$$

The off-diagonal components $\varsigma_{z\varphi}$ and $\varsigma_{\varphi z}$ (Fig. 2 b, c) and the circumferential diagonal component $\varsigma_{\varphi\varphi}$ (Fig. 2 d) arose from cross sectional magnetization process ($h_\varphi \rightarrow m_z$ and $h_z \rightarrow m_\varphi$) [21-23, 75-78].

From the point of view of industrial applications low hysteretic GMI effect with linear magnetic field dependence of the output signal are desirable [8]. Anti-symmetrical magnetic field dependence of the output voltage with linear region has been obtained for pulsed GMI effect based on detection of the off-diagonal GMI component of amorphous wires [8, 21-23]. Such pulsed scheme for GMI measurements resulted quite useful for real GMI sensors development [21-23].

As mentioned above, the shape of magnetic field dependence of the GMI effect (including off-diagonal components) is intrinsically related with the magnetic anisotropy and peculiar surface domain structure of amorphous wires [20-23]. Magnetic anisotropy of amorphous microwires in the absence of magnetocrystalline anisotropy is determined mostly by the magnetoelastic term [8, 18]. Therefore the magnetic anisotropy can be tailored by thermal treatment [8, 18, 77-78]. On the other hand recently considerable GMI hysteresis has been observed and analyzed in microwires [79]. This GMI hysteresis has been explained through the helical magnetic anisotropy [79].

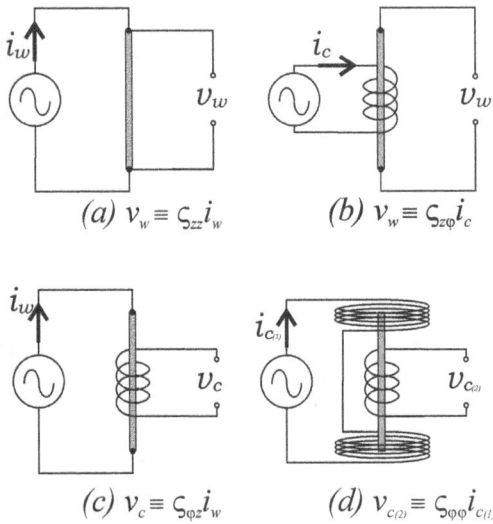

*(a)* $v_w \equiv \varsigma_{zz} i_w$   *(b)* $v_w \equiv \varsigma_{z\varphi} i_c$

*(c)* $v_c \equiv \varsigma_{\varphi z} i_w$   *(d)* $v_{c(2)} \equiv \varsigma_{\varphi\varphi} i_{c(1)}$

**Figure 2.** Methods for revealing the impedance matrix elements: (a) $\varsigma_{zz}$, (b) $\varsigma_{z\varphi}$, (c) $\varsigma_{\varphi z}$, (d) $\varsigma_{\varphi\varphi}$. Reprinted with permission from [22], A. Zhukov, M. Ipatov, V. Zhukova, C. García, J. Gonzalez, and J. M. Blanco, Development of ultra-thin glass-coated amorphous microwires for HF magnetic sensor applications, *Phys. Stat. Sol.* (a) 205, No. 6, 1367–1372 (2008) / DOI 10.1002/pssa.200778133, Copyright © 2008 with permission from WILEY (Fig. 1).

Below we overview few recent results on GMI effect in microwires paying attention on it suitability for magnetic sensors applications.

## 2.1. Frequency Dependence of GMI Effect

According to the usual definition, the complex impedance of a linear electronic element at the circular frequency *w* is given by:

$$Z(w) = Uac \,/\, Iac = R + i\,X, \qquad (8)$$

where $I$ac is the harmonic current with frequency $w$ flowing through the element and $U$ac is the harmonic voltage of the same frequency, measured between its terminals [59].

In fact as has been already mentioned [see for example 59], the definition (8) is not fully applicable to ferromagnetic conductors because usually such materials are not linear. This occurs because $U$ac is generally not proportional to $I$ac and it is not a harmonic function of time (it contains higher order harmonics) [59].

Additionally, although widely used, the definition of the GMI ratio $\Delta Z/Z$, mentioned above, may be useful for quantifying the huge attained variations of impedance, but the information about the phase shift is lost; it depends on the ambiguously chosen $H_0$ max (although the sample might be apparently magnetically saturated it does not mean that GMI is also saturated). But in any case, most of published papers use $\Delta Z/Z$ or longitudinal impedance change when studying the frequency dependence of GMI effect.

Magnetic field, H, dependence of real part, $Z_I$ of the longitudinal wire impedance $Z_{zz}$ ($Z_{zz}=Z_1+iZ_2$), measured up to 4 GHz in $Co_{66}Cr_{3.5}Fe_{3.5}B_{16}Si_{11}$ and $Co_{67}Fe_{3.85}Ni_{1.45}B_{11.5}Si_{14.5}Mo_{1.7}$ microwires are shown in Fig.3. General features of these dependences are existence of two maximums that shift to higher magnetic fields with increasing the frequency, $f$. Considerable GMI effect has been observed even at GHz- range frequencies.

On the other hand, if the maximum applied magnetic field is not high enough, impedance change induced by applied magnetic field at high frequencies decreases starting from some frequency. For example, most of the microwires show the highest GMI ratio at frequencies between 100 and 300 MHz (see Figs. 3, 4). Another interesting features observed in Figs 4a,b are that the frequency dependence of maximum GMI ratio, $\Delta Z/Z_m(f)$, measured in microwires of the same composition and different diameters presents an optimum frequency (at which $\Delta Z/Z_m$ versus $f$ exhibits the maximum)at different frequencies. Thus, for metallic nucleus diameters ranging between 8.5 and 9.0 μm the optimal frequency is about 100 MHz, while for microwires with metallic nucleus diameters between 9 and 11.7 μm the optimal frequency is about 200 MHz.

29

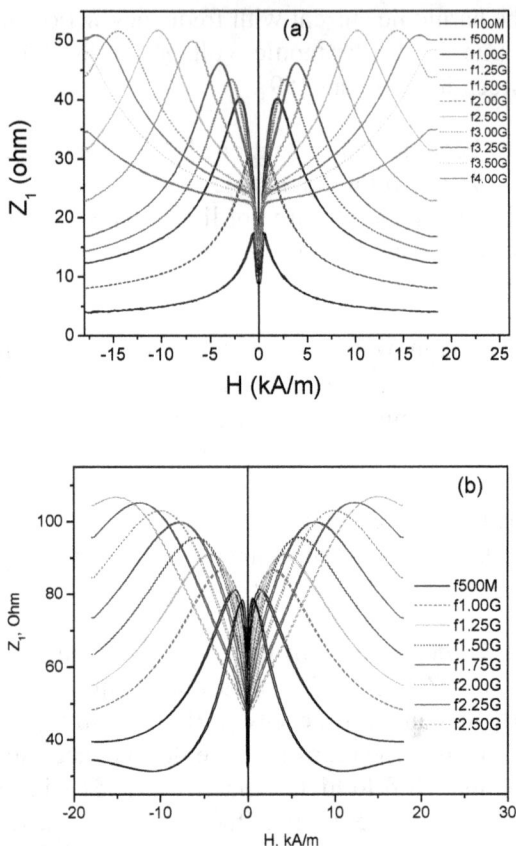

**Figure 3.** $Z_1(H)$ dependence of $Co_{66}Cr_{3.5}Fe_{3.5}B_{16}Si_{11}$ (a) and $Co_{67}Fe_{3.85}Ni_{1.45}B_{11.5}Si_{14.5}Mo_{1.7}$ (b) microwires measured at different frequencies. Reprinted with permission from [78], A. Zhukov, M. Ipatov and V. Zhukova, Amorphous microwires with enhanced magnetic softness and GMI characteristics, EPJ Web of Conferences, **29** 00052 (2012) DOI: 10.1051/ Owned by the authors, published by EDP Sciences, 2012, (Fig. 1).

As regarding to the origin of the frequency dependence of $H_m$, observed in Fig. 3 for both microwires, there are different points of view. Experimentally has been observed that the magnetic field at which the maximum occurs considerable increase with frequency, $f$ (Fig. 5).

One possible explanation for this is that the magnetic structure and the anisotropy can be different near the surface. At higher frequencies the current flows closer to the surface, then the effective anisotropy field and dispersion can change with frequency. Another reason might be connected with the frequency dependence of the domain wall permeability and aforementioned FMR contribution at higher frequencies.

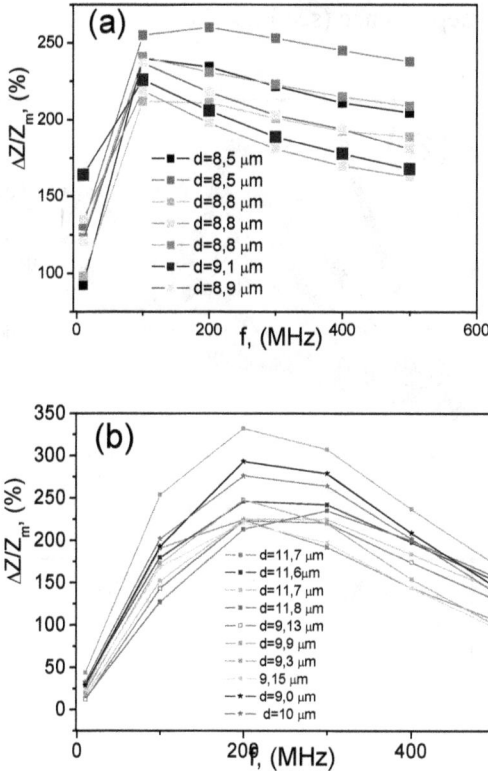

**Figure 4.** Frequency dependence of $\Delta Z/Z_m$ in $Co_{66.87}Fe_{3.66}C_{0.98}Si_{11.47}B_{13.36}Mo_{1.52}$ microwires with different metallic nucleus diameters. Reprinted with permission from [23], A. Zhukov, M. Ipatov, M. Churyukanova, S. Kaloshkin, V. Zhukova, Giant magnetoimpedance in thin amorphous wires: From manipulation of magnetic field dependence to industrial applications, *J. Alloys Comp.*, 586 (2014), S279–S286 Copyright (2014), with permission from Elsevier (Fig. 5).

The close analogy between the giant magnetoimpedance and ferromagnetic resonance has previously reported elsewhere [60, 80]. Indeed Saturation magnetization can be estimated from the equation:

$$Ms= 0.805\ 10^{-9}\ df_0^2/dH, \qquad (8)$$

where $f$ is the resonant frequency, $H$ is the applied magnetic field, $M_s$ is the saturation magnetization. This approach predicts the linear relation between the square of the resonance frequency and the applied field, $f_0^2(H)$. The experimental data for GMI effect measure in $Co_{67.05}Fe_{3.85}Ni_{1.4}B_{11.33}Si_{14.47}Mo_{1.69}$ microwires fits well with the predicted linear dependence (see Figs. 5).

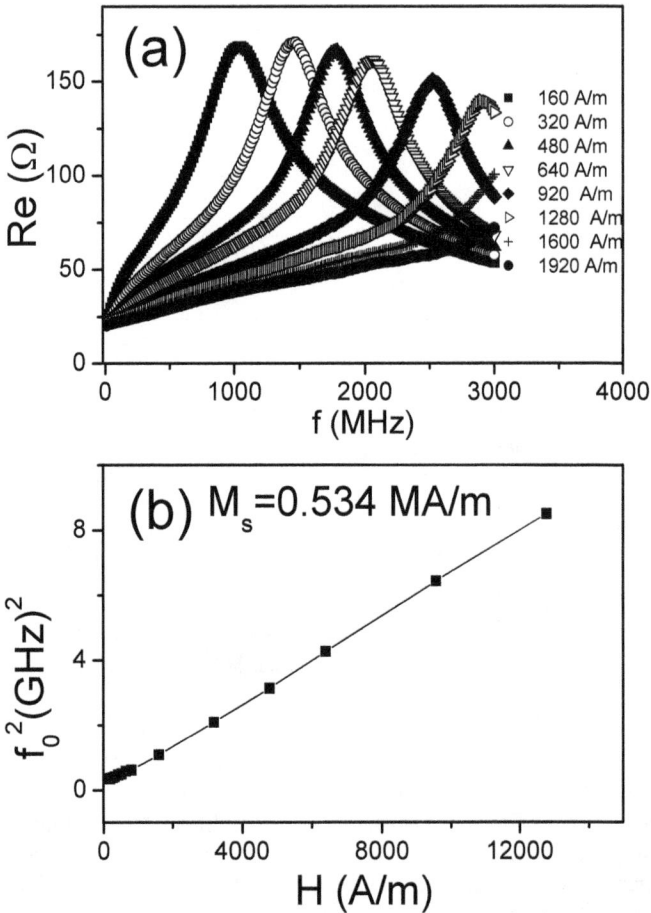

**Figure 5.** Frequency dependence of GMI effect (a) and $f_0^2(H)$ dependence (b) measured for $Co_{67.05}Fe_{3.85}Ni_{1.4}B_{11.33}Si_{14.47}Mo_{1.69}$ microwires with $d \approx 16.2$ μm, $\rho \approx 0.7$. Reprinted with permission from [80], (a)C. García, A. Zhukov, V. Zhukova, M. Ipatov, J.M. Blanco and J. Gonzalez, Effect of Tensile Stresses on GMI of Co-rich Amorphous Microwires, *IEEE Trans Magn.*, 41: 3688-3690, 2005, Copyright (2005) IEEE, (Fig. 4).

The saturation magnetization values, obtained from (8) give us quite reasonable values of about 0.5 - 0.53 MA/m. Consequently frequency dependence of GMI effect at GHz frequencies fits well with ferromagnetic resonance behavior, as previously reported elsewhere [60, 80].

Considerable GMI effect at GHz- range frequencies in observed in microwires is useful for engineering of tunable metamaterials where the GMI effect in microwires embedded into dielectric matrix is used to control the effective electromagnetic properties of composite material [8].

## 2.2. Effect of Magnetoelastic Anisotropy and GMI Hysteresis

Off-diagonal and diagonal components of GMI, measured in $Co_{67}Fe_{3.85}Ni_{1.45}B_{11.5}Si_{14.5}Mo_{1.7}$ microwires are shown in Fig. 6. As can be appreciated from Fig. 6, considerable hysteresis for both off-diagonal and longitudinal impedance is observed for studied microwires. It is important, that the GMI hysteresis does not depend on frequency: increasing the frequency the GMI hysteresis persists (Fig. 6b).

Our recent studies reveal that the magnetoelastic anisotropy has tensor character with considerable deviation of the anisotropy easy axis from transversal direction [79]. Consequently we explained the nature of observed low field hysteresis on $Z_l(H)$ and $Z_{\phi z}(H)$ (Figs. 6a and 6b) considering the existence of helical magnetic anisotropy reflected as the deviation of the anisotropy easy magnetization axis from the transversal direction [79]. Application of the circular bias magnetic field $H_B$ produced by DC current $I_B$ running through the wire affects the hysteresis and asymmetry of the MI dependence, suppressing this hysteresis when $I_B$ is high enough (see Fig. 7, where effect of bias voltage on diagonal impedance, $Z_l$, and on $S_{21}$ parameter, proportional to off-diagonal GMI component are shown).

As mentioned above, the internal stresses, $\sigma_i$, arising during simultaneous rapid quenching of metallic nucleus surrounding by the glass coating are the source of additionally magnetoelastic anisotropy. The strength of such internal stresses can be controlled by the $\rho$–ratio: strength of internal stresses increases decreasing ρ-ratio (i.e. increases with increasing of the glass volume) [45-47].

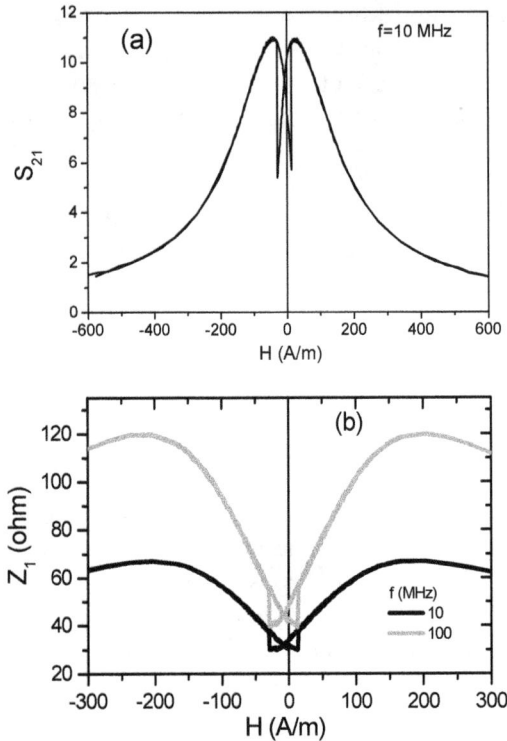

**Figure 6.** Magnetic field dependences of the coefficient $S_{21}$ at 10 MHz (a) and $Z_1(H)$ dependences at different frequencies (b) measured in $Co_{66}Cr_{3.5}Fe_{3.5}B_{16}Si_{11}$ microwire. Reprinted with permission from [78], A. Zhukov, M. Ipatov and V. Zhukova, Amorphous microwires with enhanced magnetic softness and GMI characteristics, EPJ Web of Conferences 29 00052 (2012) DOI: 10.1051/ Owned by the authors, published by EDP Sciences, 2012 (Fig. 2).

It is worth mentioning, that usually the DC magnetic field that corresponds to the maximum GMI ratio, $H_m$, is attributed to the static magnetic anisotropy field, $H_k$. Consequently, the parameter $\rho$ must be considered as one of the factors that affect both soft magnetic properties and GMI.

Fig. 8 shows the influence of the $\rho$–ratio on hysteresis loops and magnetic anisotropy field of $Co_{67.1}Fe_{3.8}Ni_{1.4}Si_{14.5}B_{11.5}$ microwires with the same composition of the metallic nucleus, but different $\rho$–ratio.

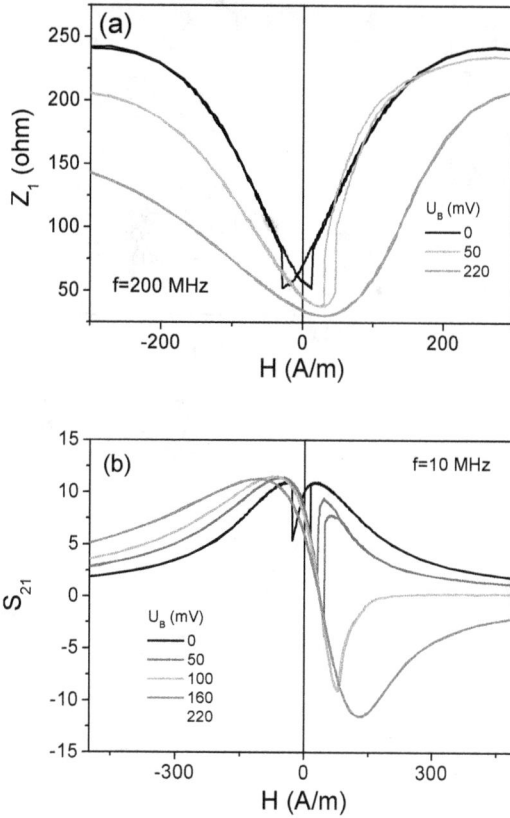

**Figure 7.** Effect of bias voltage $U_B$ on magnetic field dependence of diagonal impedance measured at 200 MHz (a) and $S_{21}$ parameter measured at 10 MHz (b) of $Co_{67}Fe_{3.85}Ni_{1.45} B_{11.5}Si_{14.5}Mo_{1.7}$ microwire. Reprinted with permission from [78], A. Zhukov, M. Ipatov and V. Zhukova, Amorphous microwires with enhanced magnetic softness and GMI characteristics, EPJ Web of Conferences 29 00052 (2012) DOI: 10.1051/ Owned by the authors, published by EDP Sciences, 2012 (Fig. 5).

As can be appreciated, Co-rich microwire with appropriate geometry and composition present excellent magnetically soft properties with low coercivities (between 4 and 10A/m) [8]. Magnetic anisotropy field, $H_k$, is found to be determined by the $\rho$–ratio, decreasing with $\rho$ (Fig. 8b), as previously reported [8].

Consequently, one can expect that the $\rho$–ratio must affect the GMI effect of studied samples.

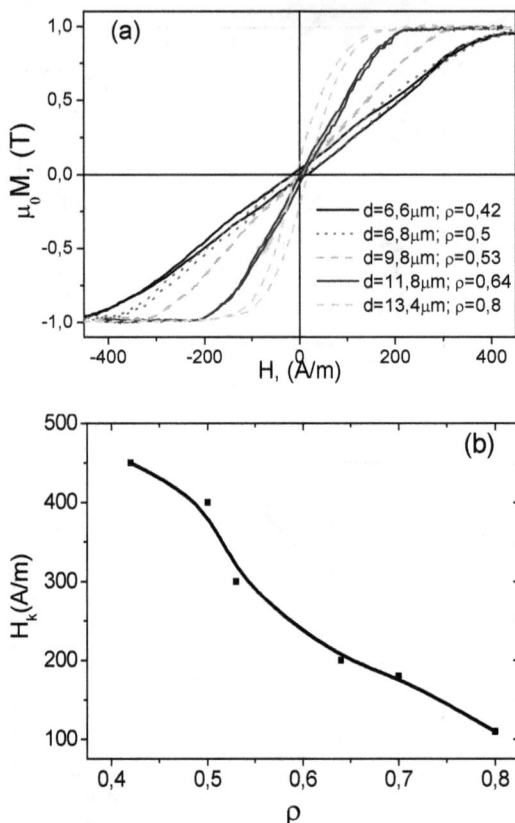

**Figure 8.** Hysteresis loops of $Co_{67.1}Fe_{3.8}Ni_{1.4}$ $Si_{14.5}$ $B_{11.5}$ $Mo_{1.7}$ microwires with different geometry (a) and dependence of $H_k$ on $\rho$- ratio (b). Reprinted with permission from [78], A. Zhukov, M. Ipatov and V. Zhukova, Amorphous microwires with enhanced magnetic softness and GMI characteristics, EPJ Web of Conferences, 29, 00052 (2012) DOI: 10.1051/ Owned by the authors, published by EDP Sciences, 2012 (Fig. 4).

Fig. 9 presents results on magnetic field dependences of $\Delta Z/Z$ measured in $Co_{67.05}Fe_{3.85}Ni_{1.4}B_{11.33}Si_{14.47}Mo_{1.69}$ microwire samples with different $\rho$–ratios.

Indeed both maximum values of the GMI ratio, $\Delta Z/Z_m$, and the magnetic anisotropy field present considerable dependence on samples geometry. It is worth mentioning, that for microwires with lowest glass coating thickness (largest $\rho$–ratio) $\Delta Z/Z_m \approx$ 600 % has been observed [8, 54].

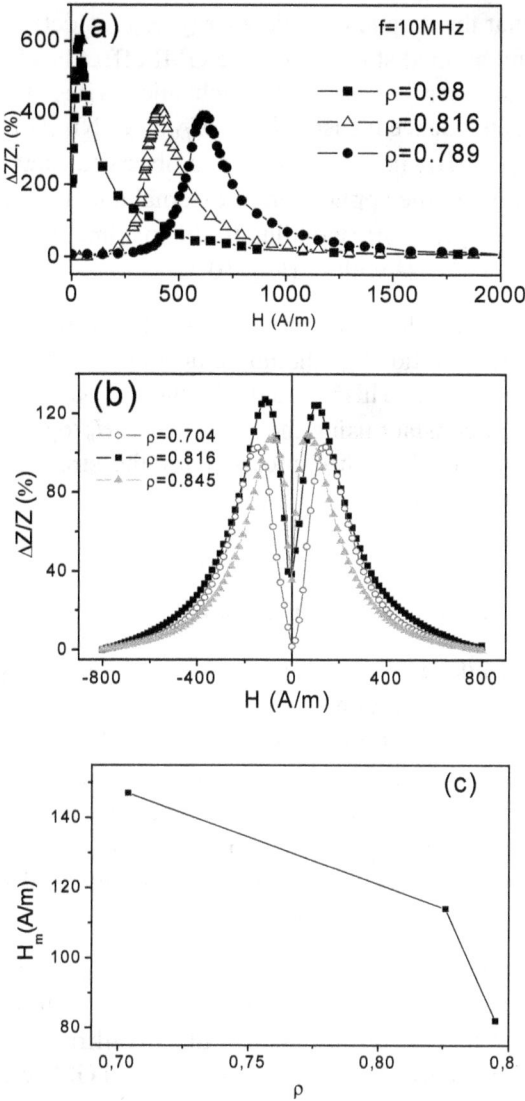

**Figure 9.** Effect of $\rho$–ratio on GMI effect in $Co_{67.05}Fe_{3.85}Ni_{1.4}B_{11.33}Si_{14.47}Mo_{1.69}$ microwire samples with different $\rho$–ratios (a, b) and dependence of field of maximums on $\rho$–ratio (c). Reprinted with permission from Ref. [54] V. Zhukova et al., Optimization of Giant Magnetoimpedance in Co-Rich Amorphous Microwires, *IEEE Trans. Magn.*, Vol. 38, 5, Part I, 2002, pp. 3090-3092. Copyright (2002) with permission from IEEE (Fig. 4) and ref [23] A. Zhukov, M. Ipatov, M. Churyukanova, S. Kaloshkin, V. Zhukova, Giant magnetoimpedance in thin amorphous wires: From manipulation of magnetic field dependence to industrial applications, *J. Alloys Comp.*, 586 (2014) S279–S286. Copyright (2014) with permission from Elsevier (Fig. 4a).

Considering that the magnetoelastic energy, $K_{me}$, is determined by both internal, $\sigma_i$, and applied stresses, $\sigma_a$, the GMI effect has been measured under tensile stresses in various Co-rich microwires. Fig. 10 present tensile stress dependence measured in $Co_{67.05}Fe_{3.85}Ni_{1.4}B_{11.33}Si_{14.47}Mo_{1.69}$ and $Co_{68.5}Mn_{6.5}Si_{10}B_{15}$ microwires. It was observed, that $\Delta Z/Z$ and $H_m$ are quite sensitive to the application of external tensile stresses, $\sigma_a$: here the magnetic field, $H_m$ corresponding to the maximum of $(\Delta Z/Z)$ shows a roughly linear increase with $\sigma$ (Fig. 10 c).

As mentioned above, the value of the *dc* axial field that corresponds to the maximum GMI ratio, $H_m$, should be attributed to the static circular anisotropy field, $H_k$. This argument allows us to estimate the magnetostriction constant using the dependence $H_m(\sigma)$ presented in the Fig. 8c and the well known expression for the stress dependence of anisotropy field [8, 54], given by

$$\lambda_s = (\mu_o M_s/3)(dH_k/d\sigma), \qquad (9)$$

where $\mu_o M_s$ is the saturation magnetization.

The $H_m(\sigma)$ dependence (see the Fig. 10 c) exhibit a slope of around 0.7 A/(m×MPa), that allows an estimation of the unstressed value of the saturation magnetostriction constant, $\lambda_{s,0}$. Estimated value of $\lambda_{s,0} \approx -2\times10^{-7}$ have been obtained, which are rather reasonable in comparison with the reported values measured from the stress dependence of initial magnetic susceptibility ($\lambda_{s,0} \approx -3\times10^{-7}$ for such composition) and the magnetostriction values measured in amorphous wires of similar compositions [54].

In fact the tendency on the change of $H_m$ under application of tensile stresses (Fig. 10 c) and the change of $H_m$ and $H_k$ with decreasing the $\rho$–ratio (Figs. 8, 9) is the same that confirms the effect of magnetoelastic anisotropy on hysteresis loops and GMI effect.

It is worth mentioning, that in some cases the value of maximum GMI ratio $(\Delta Z/Z_m)$ has a non-monotonic dependence on $\sigma$ (either applied or internal). Particularly in $Co_{68.5}Mn_{6.5}Si_{10}B_{15}$ microwire a broad maximum at around 60 MPa (with $(\Delta Z/Z_m) \approx 130$ %) (Fig. 10 c) has been observed. Similarly a maximum of $(\Delta Z/Z_m)$ at intermediate $\rho$–ratio values has been observed for $Co_{67.05}Fe_{3.85}Ni_{1.4}B_{11.33}Si_{14.47}Mo_{1.69}$ microwire (Fig. 9 b). But in the other cases monotonic decay of $(\Delta Z/Z_m)$ increasing either internal, $\sigma_i$, or applied stresses, $\sigma_a$, has been observed (Figs. 9 a and 10 a).

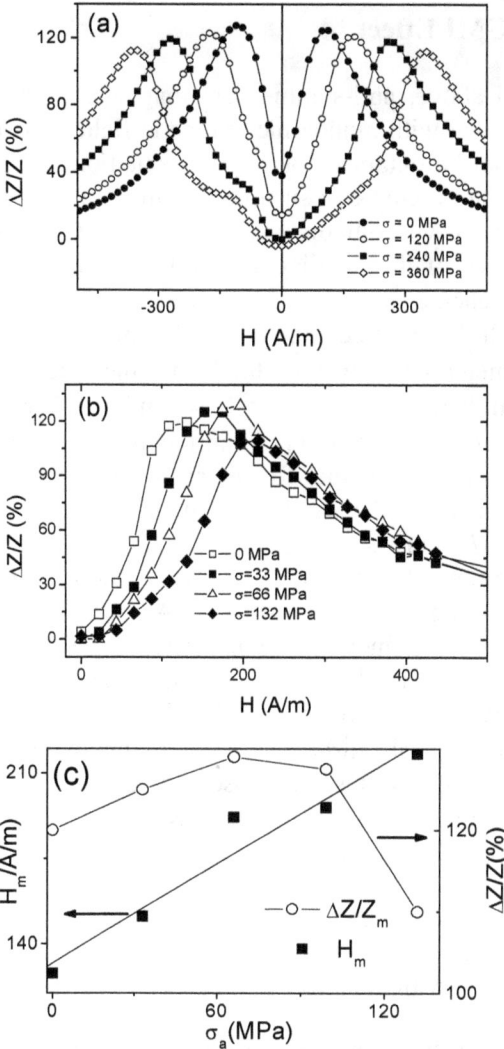

**Figure 10.** *ΔZ/Z(H)* dependences of $Co_{67.05}Fe_{3.85}Ni_{1.4}B_{11.33}Si_{14.47}Mo_{1.69}$ (a) and $Co_{68.5}Mn_{6.5}Si_{10}B_{15}$ (b) amorphous microwires measured at different $\sigma_a$ and $H_m(\sigma)$ dependence for $Co_{68.5}Mn_{6.5}Si_{10}B_{15}$ amorphous microwire (c). Reprinted with permission from Ref. [23] A. Zhukov, M. Ipatov, M. Churyukanova, S. Kaloshkin, V. Zhukova, Giant magnetoimpedance in thin amorphous wires: From manipulation of magnetic field dependence to industrial applications, *J. Alloys Comp.*, 586 (2014) S279–S286. Copyright (2001) (fig4b) and Ref. [58] A. F. Cobeño, A. Zhukov, J. M. Blanco and J. Gonzalez, Giant magneto-impedance effect in CoMnSiB amorphous microwires, *J. Magn. Magn. Mat.*, 234 (2001) L359-L365. Copyright (2001), with permission from Elsevier (Fig. 4).

## 2.3. Pulsed GMI Effect

As mentioned above, anti-symmetrical magnetic field dependence of the output voltage with monotonic magnetic field dependence at zero-field region can be obtained for pulsed excitation regime. The output voltage from the pick-up coil surrounding microwire is proportional to the off-diagonal GMI component of amorphous wires. Fig. 11 shows field dependence of the off-diagonal voltage response, $V_{out}$ measured using pulsed scheme, in $Co_{67.1}Fe_{3.8}Ni_{1.4}Si_{14.5}B_{11.5}Mo_{1.7}$ $(\lambda_s \approx -3 \times 10^{-7})$ microwire with different geometry: metallic nucleus diameter of 6 μm and total diameter 10.2 μm $(\rho \approx 0.59)$ and metallic nucleus diameter 7 μm and total diameter 11 μm $(\rho \approx 0.64)$ and of 8.2 μm total diameter of 13.7 μm $(\rho \approx 0.6)$. Observed anti-symmetrical magnetic field dependence can be suitable for determination the magnetic field direction in real sensor devices [8, 18]. It should be noted from Fig. 11 that the $V_{out}(H)$ curves exhibit nearly linear growth within the field range from $-H_m$ to $H_m$. The $H_m$ (about 240 A/m) limits the working range of MI sensor. As can be observed the amplitude of the $V_{out}$ increases with increasing the metallic nucleus diameter, $d$. On the other hand the $H_m$ value depends on the microwires geometry. Fig. 12 shows $V_{out}(H)$ dependences measured for $Co_{67.71}Fe_{4.28}Ni_{1.57}B_{12.4}Si_{11.24}Mo_{1.25}C_{1.55}$ microwires with similar d-values but with different ρ–ratios. As observed in Fig. 12b, $H_m$ decrease increasing ρ–ratios. $H_m$ should be associated with the magnetic anisotropy field.

Like in the case of conventional GMI effect, the effect of the ρ– ratio on $V_{out}$ (H) (Fig. 12 b) should be attributed to the magnetoelastic anisotropy related with the internal stresses.

Since the magnetostriction constant given by (9) is mostly determined by the chemical composition and achieves almost nearly-zero values in amorphous alloys based on Fe-Co with Co/Fe $\approx 70/5$ $\lambda_s \approx 0$ [6,11], the effect of the internal stresses determined by the ρ-ratio must be taken into account.

Additionally to the composition and geometrical factors, the magnetoelastic anisotropy can be also be tailored by reducing the internal stresses through the application of heat treatment.

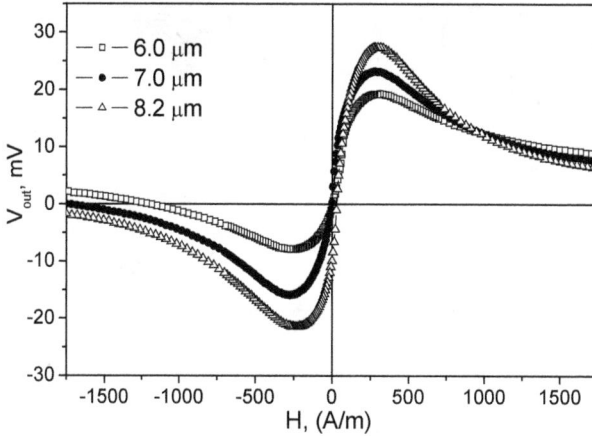

**Figure 11.** $V_{out}(H)$ response of $Co_{67.1}Fe_{3.8}Ni_{1.4}Si_{14.5}B_{11.5}Mo_{1.7}$ microwires with different diameters, $d$, and $\rho$-ratios. Reprinted with permission from [22], A. Zhukov, M. Ipatov, V. Zhukova, C. García, J. Gonzalez, and J. M. Blanco, Development of ultra-thin glass-coated amorphous microwires for HF magnetic sensor applications, *Phys. Stat. Sol.* (a) 205, No. 6, 1367–1372 (2008) / DOI 10.1002/pssa.200778133). Copyright (2008), with permission from Elsevier (Fig. 5).

In fact in pulsed exciting scheme when the sharp pulses with pulse edge time about 5 ns are produced by passing square wave multi-vibrator pulses through the differentiating circuit, overall pulsed current contains a DC component that produces bias circular magnetic field [8]. In this way low field hysteresis observed in conventional excitation scheme (Fig. 6) can be surpassed selecting adequate pulse amplitude.

On the other hand DC and even pulsed current can significantly affect the off- diagonal MI curves owing to the samples heating [8, 81]. The example presented in Fig. 13 for $Co_{67.1}Fe_{3.8}Ni_{1.4}Si_{14.5}B_{11.5}Mo_{1.7}$ microwires with vanishing magnetostriction constant shows, that the voltage pulse with the peak value above some critical affects the $V_{out}(H)$ dependence reducing the linear region. Under DC current annealing the $H_m$ decreases from 480 A/m in as-cast state to 230 A/m after 5 min annealing with 50 mA current (see Fig. 14). After the Joule heating treatment the $V_{out}(H)$ curve becomes sharper giving the higher magnetic field sensitivity and showing lower maximum field, $H_m$, related with magnetic anisotropy field. The Joule heating of nearly zero magnetostriction microwire results in decreasing of magnetoelastic energy and increasing the magnetic softness [8, 81].

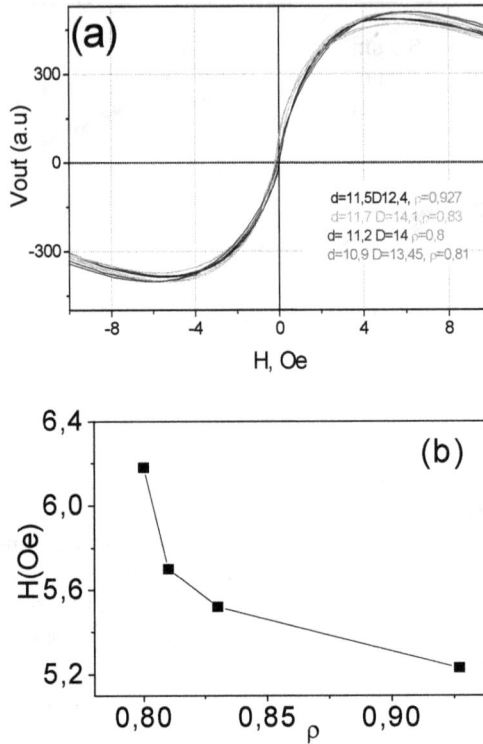

**Figure 12.** $V_{out}(H)$ response of $Co_{67.71}Fe_{4.28}Ni_{1.57}B_{12.4}Si_{11.24}Mo_{1.25}$ $C_{1.55}$ with similar metallic nucleus diameters, $d$, and different $\rho$-ratios (a) and $H(\rho)$ dependence (b). Reprinted with permission from Ref. [23] A. Zhukov, M. Ipatov, M. Churyukanova, S. Kaloshkin, V. Zhukova, Giant magnetoimpedance in thin amorphous wires: From manipulation of magnetic field dependence to industrial applications, *J. Alloys Comp.*, 586 (2014), S279–S286 Copyright (2014), with permission from Elsevier (Fig. 7).

Similarly in high negative magnetostriction microwire $Co_{74}B_{13}Si_{11}C_2$ $(\lambda_s \approx - 10^{-6})$ with metallic diameter of 10 μm strongly hysteretic off-diagonal MI curve in as-prepared state has been observed [81]. DC Joule heating affects the character of the $V_{out}(H)$ dependence of $Co_{74}B_{13}Si_{11}C_2$ microwire: the hysteretic MI curve transforms into the unhysteretic one with large enough nearly-linear region (see Fig. 15). Low sensitivity of negative magnetostriction microwires should be attributed to high enough magnetoelastic energy, related with high negative magnetostriction and stresses induced in metallic nucleus by the glass coating during simultaneous quenching. Joule heating reduces internal stresses and enhances the $V_{out}$.

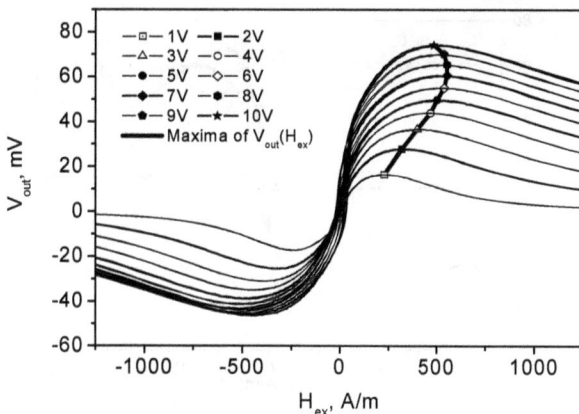

**Figure 13.** Effect of pulse annealing with different pulse amplitude on off diagonal GMI of $Co_{67.1}Fe_{3.8}Ni_{1.4}Si_{14.5}B_{11.5}Mo_{1.7}$ microwires. Reprinted with permission from Ref. [81]. V. Zhukova, M. Ipatov, J. González, J. M. Blanco and A. P. Zhukov, Development of Thin Microwires With Enhanced Magnetic Softness and GMI, *IEEE Trans. Magn.*, Vol. 44, No. 11, Part 2 pages 3958-3961, November 2008, Copyright (2008), with permission from IEEE (Fig. 5).

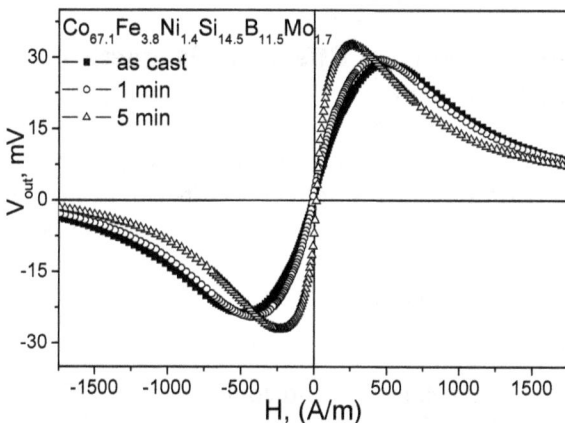

**Figure 14.** Field dependence of the off-diagonal voltage response of $Co_{67}Fe_{3.85}Ni_{1.45}B_{11.5}Si_{14.5}Mo_{1.7}$ Joule-heated microwire annealed with 50 mA currents for different time. Reprinted with permission from Ref. [81] V. Zhukova, M. Ipatov, J. González, J. M. Blanco and A. P. Zhukov, Development of Thin Microwires with Enhanced Magnetic Softness and GMI, *IEEE Trans. Magn.*, Vol. 44, No. 11, Part 2 pages 3958-3961, November 2008, Copyright (2008), with permission from IEEE (Fig. 6).

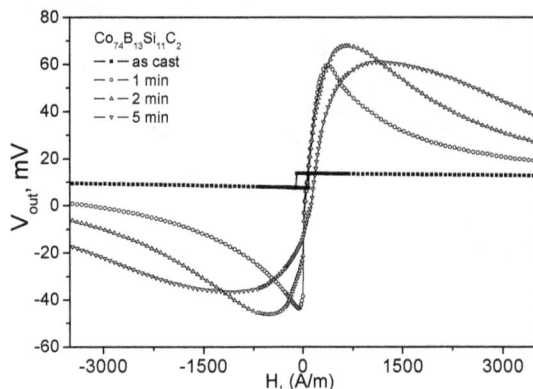

**Figure 15.** Field dependence of the off-diagonal voltage response of $Co_{74}B_{13}Si_{11}C_2$ microwire in as-prepared state and annealed with 50 mA currents for different time. Reprinted with permission from Ref. [81] V. Zhukova, M. Ipatov, J. González, J. M. Blanco and A. P. Zhukov, Development of Thin Microwires with Enhanced Magnetic Softness and GMI, *IEEE Trans. Magn.,* Vol. 44, No. 11, Part 2 pages 3958-3961, November 2008, Copyright (2008), with permission from IEEE (Fig. 7).

## 2.4. Manipulation of GMI Effect and Magnetic Properties by Heat Treatments

The aforementioned results show the possibility to tailor the microwire magnetic properties and GMI effect for its application in magnetic sensors through the selection of their composition and/or thermal treatment conditions. One can see that the current annealing with 50 mA DC current reduces the $H_m$ from 480 A/m in as-cast state to 240 A/m after 5 min annealing (Fig. 14).

Similarly, current annealing (due to Joule heating) induced changes in GMI ratio (Fig. 16). This effect should be mostly attributed to the stress relaxation (although electrical current also induce circular magnetic field).

It was previously demonstrated that application of stress and/or magnetic field during annealing of amorphous materials may induce strong additional magnetic anisotropy. In the case of microwires this induced anisotropy can be reinforced due to strong internal stresses (therefore even conventional annealing must be considered as stress-annealing) [8, 18, 81, 82]. In some cases this results in drastic changes of hysteretic magnetic properties and GMI behavior [8, 18, 83-85]. As an example, application of axial magnetic field during annealing induces axial magnetic anisotropy in Co-rich microwires (Fig. 17). Here hysteresis loops of $Co_{67}Fe_{3.85}Ni_{1.45}B_{11.5}Si_{14.5}Mo_{1.7}$ microwires ($d$=22.4 μm, $D$=22.8 μm) annealed by Joule heating without (CA) and under application of axial magnetic field (FCA) are shown. As can be appreciated, application of magnetic field during annealing completely resulted in the opposite tendency in changing of magnetic properties induced by annealing: increasing of remanent magnetization and decreasing of coercivity after FCA is observed, while CA treatment induced decreasing of the remanence and of the coercivity.

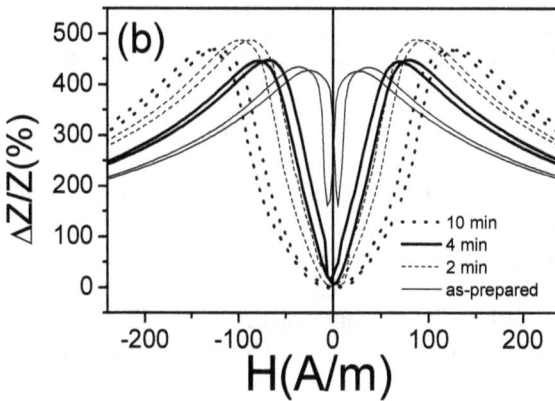

**Figure 16.** *ΔZ/Z(H)* dependences of heated $Co_{67}Fe_{3.85}Ni_{1.45}B_{11.5}Si_{14.5}Mo_{1.7}$ microwire measured at *f*=30 MHz and *I*=1 mA in microwire subjected to CA annealing at 40 mA for different time. Reprinted with permission from Ref. [8] V. Zhukova, M. Ipatov and A Zhukov, Thin Magnetically Soft Wires for Magnetic Microsensors, Sensors 9: 9216-9240, 2009 Copyright (2009), with permission from Sensors (Fig. 9 b).

In the case of Fe-rich microwires subjected to the annealing in the presence of tensile stresses complete change of magnetic anisotropy can be realized (Fig. 18).

Stress annealing of $Fe_{74}B_{13}Si_{11}C_2$ microwires resulted in induction of considerable stress induced anisotropy [83-85]. The shape of hysteresis loop completely changes and the strength of induced changes depends on time and temperature of annealing (Fig. 18 a). In this case the easy axis of magnetic anisotropy has been changed from axial to transversal [83-85]. Additionally, application of stress during measurements of stress-annealed microwires with well-defined transverse anisotropy results in drastic change of the hysteresis loop (Fig. 18 b).

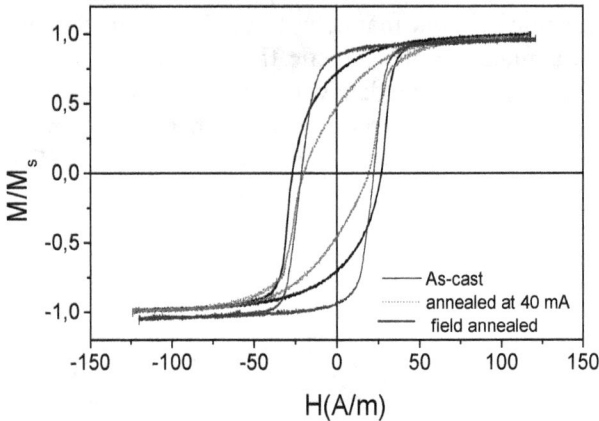

**Figure 17.** Effect of CA and FCA on bulk hysteresis loops of $Co_{67}Fe_{3.85}Ni_{1.45}B_{11.5}Si_{14.5}Mo_{1.7}$ microwires ($d$=22.4 μm, $D$=22.8 μm). Reprinted with permission from [78], A. Zhukov, M. Ipatov and V. Zhukova, Amorphous microwires with enhanced magnetic softness and GMI characteristics, EPJ Web of Conferences 29 00052 (2012) DOI: 10.1051/ Owned by the authors, published by EDP Sciences, 2012, (Fig. 7 b).

Origin of such stress-induced anisotropy is related with so-called "Back stresses" originated from the composite origin of glass-coated microwires annealed under tensile stress: compressive stresses compensate axial stress component and under these conditions transversal stress components are predominant [83-85].

Consequently, these stress annealed samples exhibit stress-impedance effect, i.e. impedance change ($\Delta Z/Z$) under applied stress, $\sigma$, observed in samples with stress induced transversal anisotropy (see Fig. 19) [83-85].

It should be assumed that the internal stresses relaxation after heat treatment should drastically change both the soft magnetic behavior and the $\Delta Z/Z(H)$ dependence due to the stress relaxation, induced magnetic anisotropy and change of the magnetostriction constant under annealing.

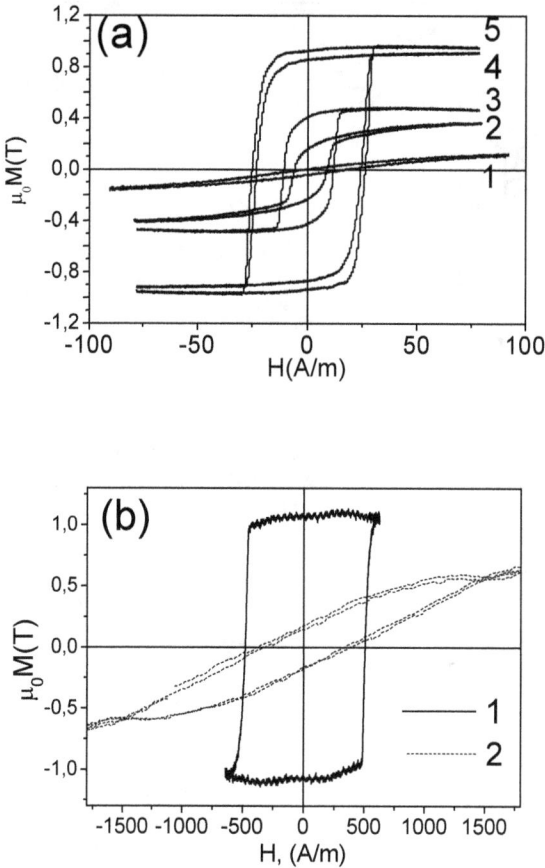

**Figure 18.** Hysteresis loops of $Fe_{74}B_{13}Si_{11}C_2$ microwire annealed under applied stress of 500 MPa (a) at (1) – 300 °C 3 hours, (2) – 280 °C 40 min, (3) – 265 °C 40 min, (4) – 235 °C 40 min and (5) – 215 °C 40 min and (b) stress induced changes of hysteresis loops of the same microwires (1- measured under applied stress, 2- measured without stress). Reprinted with permission from [84], A. Zhukov, V. Zhukova, V. Larin, J. M. Blanco and J. Gonzalez, Tailoring of magnetic anisotropy of Fe-rich microwires by stress-induced anisotropy, *Physica B*, 384 (2006) 1-4 Copyright (2006), with permission from Elsevier. (Fig. 1, 3).

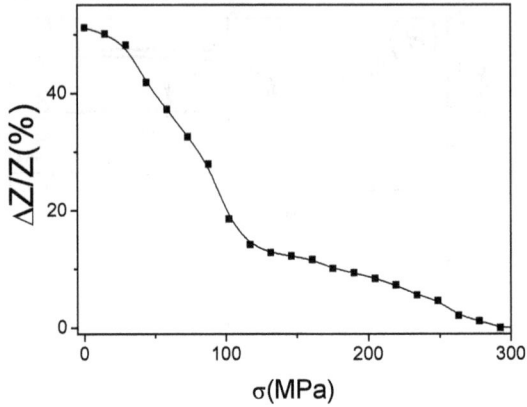

**Figure 19.** Stress impedance effect of stress annealed $Fe_{74}B_{13}Si_{11}C_2$ glass-coated microwire under stress (468 MPa) at 275°C for 0.5h measured at frequency, $f=10$ MHz for the driving current amplitude of 2 mA. Reprinted with permission from [78], A. Zhukov, M. Ipatov and V. Zhukova, Amorphous microwires with enhanced magnetic softness and GMI characteristics, EPJ Web of Conferences, 29, 00052 (2012) DOI: 10.1051/ Owned by the authors, published by EDP Sciences, 2012, (Fig. 8).

# 3. Influence of Partial Crystallization and Nanocrystallization on Magnetic Properties and GMI

Usually crystallization of amorphous materials results in considerable magnetic hardening and therefore considerable reduction or even disappearance of the GMI effect [8, 18]. But in some particular compositions, particularly in so-called Finemet-type compositions ($Fe_{73.8}Cu_1Nb_{3.1}Si_{13}B_{9.1}$) considerable magnetic softening and increasing of the GMI ratio as-compared with as-cast amorphous state have been observed [37-39, 86].

All as-prepared Finemet-type microwires present squired hysteresis loops similar to Fe-rich amorphous microwires. (Fig. 20).

The coercivity, $H_C$, of as-prepared Finemet-type microwires depends on ratio $\rho = d/D$ (Figs. 20 - 22).

Observed magnetic softening by most of researchers is explained by the fact, that the exchange correlation length of the matrix is larger than the average intergranular distance, $d$, and the exchange correlation length of the grains is larger than the grain size, $D$. Enhanced magnetic softness of Fe-rich nanocrystalline alloys was also attributed to a second complementary reason: the opposite sign of the magnetostriction constant of crystallites and residual amorphous matrix, which allows achieving of vanishing average magnetostriction constant.

Magnetic properties of as-prepared and annealed $Fe_{73.4}Cu_1Nb_{3.1}Si_xB_{22.5-x}$ ($x$ = 11.5, 13.5 and 16.5) and $Fe_{73.4-x}Cu_1Nb_{3.1}Si_{13.4+x}B_{9.1}$ ($0 \leq x \leq 1.1$) "*Finemet*"-type microwires significantly affected by their geometry [18, 39, 86]. Figs. 21 and 22 represent dependences of coercivity, $H_C$, on annealing temperature, $T_{ann}$ for both "*Finemet*"-type microwires, $Fe_{73.4}Cu_1Nb_{3.1}Si_xB_{22.5-x}$ ($x$ = 11.5, 13.5 and 16.5) and $Fe_{73.4-x}Cu_1Nb_{3.1}Si_{13.4+x}B_{9.1}$ ($0 \leq x \leq 1.1$).

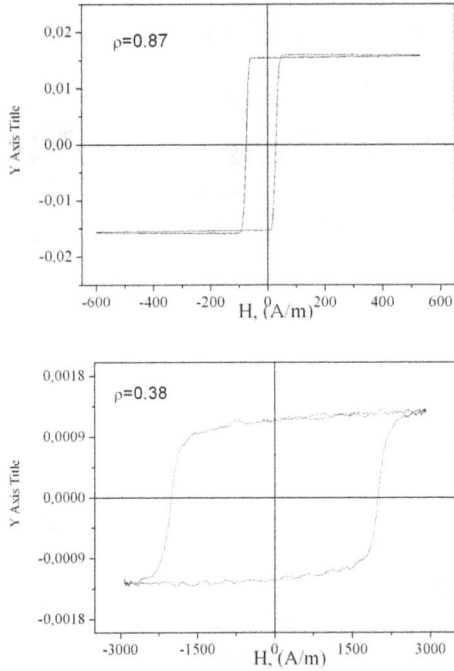

**Figure 20.** Hysteresis loops of as-prepared $Fe_{73.8}Cu_1Nb_{3.1}Si_{13}B_{9.1}$ microwires with different $\rho$=d/D ratios: a) $\rho$ =0.87, b) $\rho$ =0.38. Reprinted with permission from [86], A. Zhukov, M. Churyukanova, L. Gonzalez, A. Talaat, V. Zhukova, B. Hernando, M. Ilyn, J. Gonzalez and S. Kaloshkin, Influence of Magnetoelastic Anisotropy on Properties of nanostructured Microwires, *Advanced Materials Research*, Vol. 646 (2013), pp 59-66. Copyright (2013), with permission from Scientific.Net (Fig. 1).

Like in the case of Fe- based amorphous microwires, the coercivity of the sample $Fe_{72.3}Cu_1Nb_{3.1}Si_{14.5}B_{9.1}$ strongly increases as the ratio $\rho$ decreases.

For both compositions of *"Finemet"*-type microwires the decrease of $H_C$ has been, generally, observed at $T_{ann}$ below 673 K, which could be ascribed to the internal relaxation stresses effect. In the range of annealing temperature of 673-723 K, a weak local minimum of $H_C$ has been observed, with the temperature of that minimum depending on both alloy composition and geometry. Such a decrease of $H_C$ could be ascribed to the structural relaxation of the material remaining the amorphous character such as has been widely reported in metallic glass alloys. A small relative hardening (increase of coercivity) can be

observed after annealing around 723–773 K, which could be ascribed to
the very beginning of the first stage of devitrification [2, 18, 31-38,
86-88]. It is interesting to note that the sample's geometry affects the
value and the position of the local extremes on the $H_C(T_{ann})$ dependence
(see Figs. 21, 22). A deeper softening (optimum softness) with a rather
low value of $H_c$ is obtained in samples treated at $T_{ann}$ = 773–873 K. The
position of this minimum is also strongly affected by the sample
geometry (see Figs. 21,22). This magnetic softening is related to the
nanocrystallization process owing the precipitation of fine grains
(10–15 nm) of α-Fe(Si) phase within the amorphous matrix. Such
interpretation has been confirmed by the X-ray diffraction studies of
as-prepared and annealed at different temperatures samples [8, 86-89]
(see Fig.23). As can be deduced from Fig. 23, amorphous state was
observed for the as-prepared samples $Fe_{73.4}Cu_1Nb_{3.1}Si_xB_{22.5-x}$ with
x =11.5. The X-ray diffraction patterns for the annealed samples with
x = 11.5 indicate the appearance of a-Fe(Si) crystalline grains
(randomly textured) for annealing temperature, $T_{ann}$ above 773 K (see
Fig. 23). Annealing above 973 K causes the appearance of new boride
phases. The mean grain diameter of precipitated crystallites are
estimated using the Debye-Sherrer equation is 2; 12 and 14 nm for
$T_{ann}$= 773, 823 and 1023 K respectively [86-89].

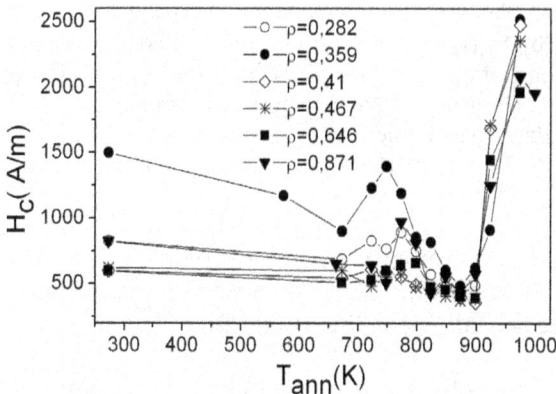

**Figure 21.** Annealing temperature dependence of coerctivity
of $Fe_{71.8}Cu_1Nb_{3.1}Si_{15}B_{9.1}$ microwires with different $\rho$ -ratios. Reprinted with
permission from [87], V. Zhukova, A. F. Cobeño, A. Zhukov, J. M. Blanco,
V. Larin and J. González, *Coercivity of glass-coated
$Fe_{73.4-x}Cu_1Nb_{3.1}Si_{13.4+x}B_{9.1}$ (0<x<1.6) microwires, Nanostructured Materials,*
11 (1999) 1319. Copyright (1996) with permission from Elsevier (Fig. 3).

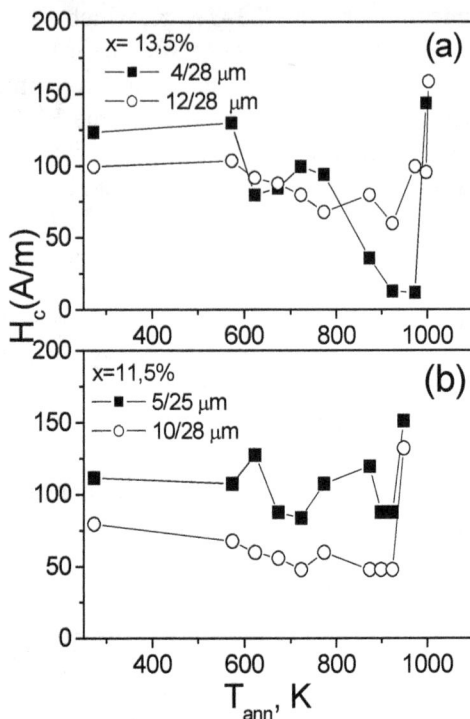

**Figure 22.** Annealing temperature dependence of coerctivity of $Fe_{73.4}Cu_1Nb_{3.1}Si_xB_{22.5-x}$ microwires with different $\rho$ -ratios. Reprinted with permission from [37] J. Arcas, C. Gómez-Polo, A. Zhukov, M. Vázquez, V. Larin and A. Hernando, "Magnetic properties of amorphous and devitrified FeSiBCuNb glass-coated microwires", Nanostructured Materials V.7 No 8. (1996) 823-834. Copyright (1996), with permission from Elsevier (Fig. 4 a, b).

Excellent magnetic softness (coercivity of about 15 A/m) has been realized in $Fe_{73.5}Cu_1Nb_3Si_xB_{22.5-x}$ microwires with $x = 13.5$ [8, 37, 86] at appropriate annealing conditions (see Fig. 21, 22).

Optimum softest behavior, that is, a deeper softening with very low value of coercivity is obtained in the samples treated 773–873 K which could be ascribed to the fact that the first crystallization process has been developed, leading to fine nanocrystals α-Fe (Si) of grain size around 10 nm with 70–80 % of relative volume, such it has been widely reported for *Finemet* ribbons (see refs. [31-38]). An abrupt increase of the coercivity is shown by the samples treated at temperature above 873 K, indicating the beginning of such increase should be connected to the precipitation of iron borides (with grain size larger than 50 nm) which implies a magnetic hardening character. The beginning of the increase of $H_c$ varies depending mainly on the sample composition as well as on the geometry.

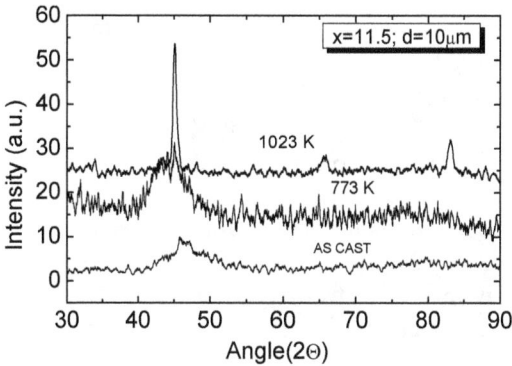

**Figure 23.** Evolution of the X-ray diffraction after annealing of $Fe_{73.4}Cu_1Nb_{3.1}Si_xB_{22.5-x}$ microwires ($x$=11.5 %), $d$=10 μm. Reprinted with permission from [37] J. Arcas, C. Gómez-Polo, A. Zhukov, M. Vázquez, V. Larin and A. Hernando, Magnetic properties of amorphous and devitrified FeSiBCuNb glass-coated microwires, *Nanostructured Materials*, V.7 No 8, (1996) 823-834 Copyright (1996), with permission from Elsevier (Fig. 1a).

Observed above effect of the thickness of glass coating has been interpreted as the consequence of both strong internal stresses and the difference in the thermal exchange conditions during fabrication and annealing due to isolating properties of the glass coating. As a consequence, the coercivity of the as-prepared and treated (573–973 K) samples results to be very sensitive to the $d/D$ ratio with significant difference in the softening (nanocrystallization) and hardening (second crystallization process) character.

It is important that the hysteresis loop in certain microwire compositions remains roughly rectangular in the whole range of annealing temperatures for certain compositions [37].

Such temperature dependence of the hysteresis loops allows one to tailor the coercivity of microwires with rectangular hysteresis loops in a wide range of coercivities. This is in fact quite important for some particular applications of glass-coated microwires, related with magnetic encoding using magnetic TAGs.

It should be indicated that $H_c(T_{ann})$ dependence is strongly affected by the microwires composition. Thus in slightly different composition $Fe_{72.3}Cu_1Nb_{3.1}Si_{14.5}B_{9.1}$ first maximum on $H_c(T_{ann})$ dependence takes place at the same range of the annealing temperature as in $Fe_{73.4}Cu_1Nb_{3.1}Si_xB_{22.5-x}$ ($x$ = 11.5 and 13.5), but observed magnetic hardening is much stronger [88]. This strong magnetic hardening of $Fe_{72.3}Cu_1Nb_{3.1}Si_{14.5}B_{9.1}$ microwire is followed by new magnetic softening with an increase of $T_{ann}$ above 823 K and consequent new magnetic hardening at $T_{ann}$ > 923 K which is accompanied by deterioration of mechanical properties [88].

In order to understand such considerable difference in $H_c(T_{ann})$ in very similar compositions X-ray analysis of $Fe_{72.3}Cu_1Nb_{3.1}Si_{14..5}B_{9.1}$ has been performed. X-ray analysis of $Fe_{72.3}Cu_1Nb_{3.1}Si_{14..5}B_{9.1}$ microwires do not detect a presence of crystalline phases in as-prepared state as well as after annealing at 823 K. On the other hand, transmission electron microscopy (TEM) diagrams of the sample annealed at 823 K allow one to detect a small amount of fine grains of $\alpha$-Fe, $\gamma$-Fe, and $\alpha$-Fe(Si) [88]. Observed structural features permit one to correlate observed magnetic hardening with precipitation of fine crystallites. Such a difference in magnetic behavior with conventional *Finemet*-type microwires can be attributed to the slightly different alloy compositions, different size of precipitating nano-grains as well as to the effect of high internal stresses. It is well known that the best magnetic softness is achieved when the nanocrystalline structure consisting of small $\alpha$-Fe(Si) grains and amorphous matrix has a vanishing magnetostriction constant. It is possible that even a small change of the alloy composition does not permit one to achieve such a vanishing magnetostriction constant. On the other hand, strong internals stresses due to glass coating can result in a change of structure of the precipitating fine grains. Thus, strong internal stresses (about 100 MPa or even more) are mainly induced by the difference in the thermal expansion coefficients of the glass and the metallic nucleus [2-4]. It is well known that internal strains of different nature can be the origin of martensite-type transformation in alloys of Fe. Probably, the strong internal stresses induce a precipitation of the $\gamma$-Fe fine grains instead or apart from $\alpha$-Fe(Si) grain during the first stage of the crystallization process. Particularly, the presence of crystalline $\gamma$-Fe could be the reason for the observed magnetic hardening [88]. As a consequence, the origin of such strong magnetic hardening at low annealing temperatures without deterioration of mechanical properties could be ascribed to some peculiarities of the first recrystallization process under the effect of strong internal stresses induced by glass coating and differences in the alloy composition.

Annealing temperature dependence of coectivity of $Fe_{71.8}Cu_1Nb_{3.1}Si_{15}B_{9.1}$ microwires shown in Figs. 21, 22 present considerable magnetic softening at annealing temperatures, $T_{ann}$, between 800 and 900 K as previously observed in other Finemet-type materials and Finemet-type microwires [18, 87].

Consequently, although GMI effect in as-prepared Fe-rich microwires is rather small, after annealing we observed increasing of the GMI effect (Fig. 24). Enhancement of the $\Delta Z/Z$ ratio is related with magnetic softening of studied microwires after annealing. Generally, measurements of the GMI effect in nanocrystalline microwires involve preparation of the electrical connections of rather brittle nanocrystalline samples. The work on optimization of the GMI ratio during nanocrystallization is in progress.

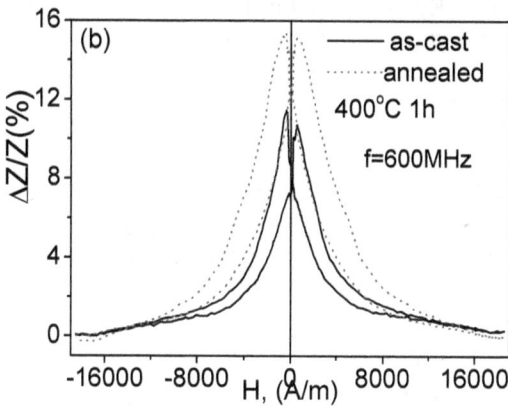

**Figure 24.** $\Delta Z/Z(H)$ dependences of $Fe_{73.8}Cu_1Nb_{3.1}Si_{13}B_{9.1}$ amorphous microwires measured in as-prepared and annealed at 400C samples at 600 MHz. Reprinted with permission from [86], A. Zhukov, M. Churyukanova, L. Gonzalez, A. Talaat, V. Zhukova, B. Hernando, M. Ilyn, J. Gonzalez and S. Kaloshkin, Influence of Magnetoelastic Anisotropy on Properties of nanostructured Microwires, *Advanced Materials Research*, Vol. 646, (2013) pp. 59-66. Copyright (2013), with permission from Scientific.Net (Fig. 4).

Typical DSC curves of as-prepared Finemet-type microwires are shown in Fig. 25.

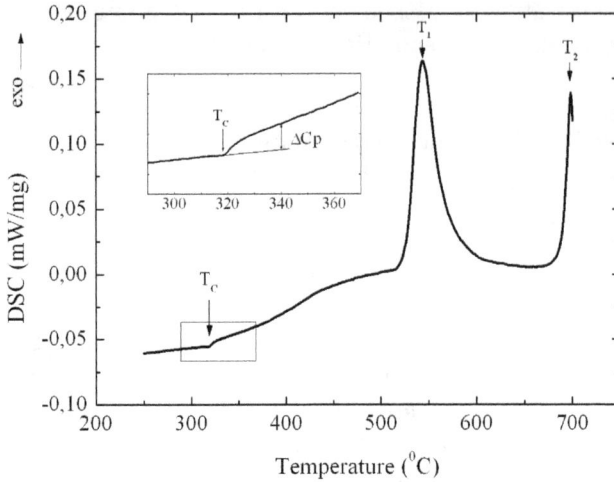

**Figure 25.** DSC curves of as-prepared $Fe_{70.8}Cu_1Nb_{3.1}Si_{14.5}B_{10.6}$ microwires with $\rho \approx 0.5$. Reprinted with permission from [86], A. Zhukov, M. Churyukanova, L. Gonzalez, A. Talaat, V. Zhukova, B. Hernando, M. Ilyn, J. Gonzalez and S. Kaloshkin, Influence of Magnetoelastic Anisotropy on Properties of nanostructured Microwires, Advanced Materials Research Vol. 646, (2013), pp 59-66. Copyright (2013), with permission from Scientific.Net (Fig. 5).

Magnetic transition at the temperature around 300 °C ($T_C$) and two peaks corresponding to two stages of the crystallization ($T_1$ and $T_2$) can be appreciated. $T_1$ must be attributed to the precipitation of $\alpha$-Fe nanocrystals, and $T_2$ – to the Fe-B phase precipitation upon annealing [86].

The calorimetric peak in vicinity of $T_C$ exhibits considerable dependence on $\rho$-ratio for the studied microwires (see Fig. 26).

For the samples with low $\rho$-ratio values (i.e. corresponding to the samples with the highest internal stresses) the DSC signal is smaller, i.e. the calorimetric peak in vicinity of $T_C$ is smaller in case of strong internal stresses.

Glass removal results in considerable change of heat capacity $\Delta Cp$ (Fig. 27). This confirms that internal stresses affect the value of the peak of heat capacity in the Curie point since glass removal releases the stresses.

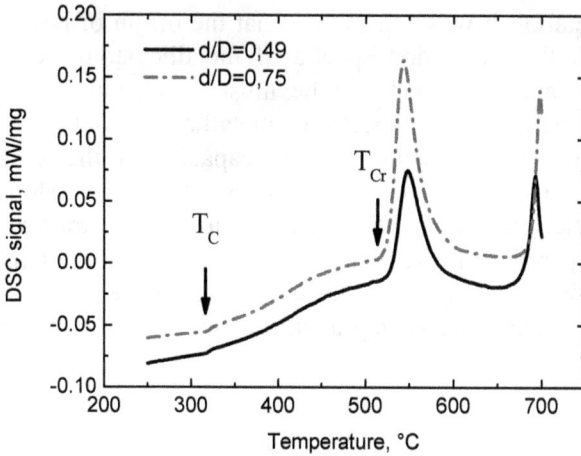

**Figure 26.** DSC curves of as-prepared $Fe_{70.8}Cu_1Nb_{3.1}Si_{14.5}B_{10.6}$ microwires with different $\rho$–ratio. Reprinted with permission from [86], A. Zhukov, M. Churyukanova, L. Gonzalez, A. Talaat, V. Zhukova, B. Hernando, M. Ilyn, J. Gonzalez and S. Kaloshkin, Influence of Magnetoelastic Anisotropy on Properties of nanostructured Microwires, *Advanced Materials Research*, Vol. 646 (2013), pp 59-66. Copyright (2013), with permission from Scientific.Net (Fig. 6).

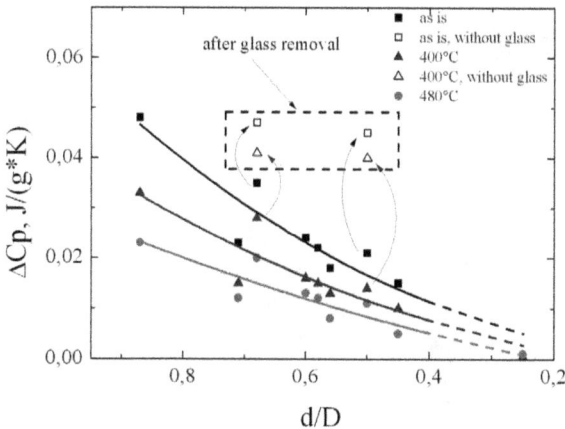

**Figure 27.** DSC curves of $Fe_{70.8}Cu_1Nb_{3.1}Si_{14.5}B_{10.6}$ microwires with $\rho \approx 0.5$ for annealing during 4 min at different temperatures. Reprinted with permission from [86], M. Churyukanova, V. Zhukova, S. Kaloshkin, A. Zhukov, Effect of magnetoelastic anisotropy on properties of Finemet-type microwires, *Journal of Alloys and Compounds*, 536S (2012) S291– S295 Copyright (2012), with permission from Elsevier (Fig. 7).

For interpretation it was considered, that the origin of heat capacity is related with the oscillation spectra of metallic atoms. Consequently negligible changes of heat can be must be attributed to negligible changes of the oscillation spectra of metallic atoms at transition from ferromagnetic to paramagnetic state capacity in the samples with highest internal stresses. This means that the elongation in ferromagnetic state related with the magnetostrictive strain is negligible low as-compared with the elongation under the effect of the stresses. Consequently this elongation does not change at the Curie point and as a result we cannot observe any alterations in heat capacity by the DSC method.

# 4. GMI-related Applications

## 4.1. Magnetic Field Detection

For magnetic field sensing presently Hall-effect sensors, Magnetoresistive sensors and Fluxgate sensor are widely applied. Quite recently Giant magneto-impedance effect (GMI) effect exhibiting extremely high magnetic field sensitivity has been observed and a novel type of magnetic sensor called GMI sensors has been recently developed and proposed [8, 9, 21-23]. Actually, most recent applications derive from the magnetoimpedance effect in the MHz frequency range observed mostly in low-magnetostrictive microwires.

First magnetic field sensor has been developed by K. Mohri and T. Uchiyama using magnetic wires with 100 μm diameter [89]. Certainly the main interest in studies of GMI effect is related to extremely high magnetic field sensitivity exhibited by the soft magnetic materials with high GMI effect. In this regard, different families of magnetic wires have been mentioned many times owing to highest GMI effect reported many times[9,16, 21-23]. Consequently magnetic wires have been proposed for a number of technological applications, and most specifically as sensing elements of various devices.

Later for the sensor miniaturization magnetic wire diameter has been drastically reduced. This has been done by using cold drawing process [16, 90] or using glass-coated microwires [9, 22, 23].

Nowadays FeCoSiB magnetoimpedance microwires are currently employed by Aichi Steels in Japan integrated in CMOS circuit as kind of magnetic compass for a number of uses in mobile systems [12, 91]. Last magnetic field MI sensor generations are characterized by quite interesting sensitivity of 1 pT [90-91].

These GMI sensors are very attractive for weak magnetic field sensing. The Table 1 below summarizes the main characteristics of different types of magnetic sensors such as Hall-effect, Magnetoresistive, Fluxgate and GMI sensors. Based upon an examination of the table and taking into account the requirements to magnetic sensor (small size, high sensitive, high-speed and low-cost sensors), it can be concluded that GMI sensor is the promising choice. The Hall Effect and Magnetoresistive do not meet the requirements of sensitivity and Fluxgate – of head size, power consumption and do not measure dc magnetic fields. Consequently, only GMI sensor simultaneously realizes highly sensitive, micro-sized, and cover the required frequency range starting from dc fields. Fig. 28 shows the noise level comparison of different magnetic sensors [91]. As one can see low- cost GMI technology already compares well with other high sensitivity magnetometers exhibiting one of the lowest noise levels between the non-cryogenic magnetometers. Another benefit of GMI technology would be the relatively high frequencies which could be sensed. In comparison, the maximal frequency of fluxgate sensor is limited to a few kHz. The drawbacks of GMI technology are comparatively large dimensions and complicity in integration into CMOS process. Further we will consider the GMI sensors.

As mentioned above, in sensor application, the pulse excitation is preferred because of simple electronic design and low power consumption. Additionally low hysteretic GMI effect with linear magnetic field dependence of the output signal is desirable [8]. Abovementioned pulsed excitation scheme allowed to obtain anti-symmetrical magnetic field dependence of the output voltage with linear region. This method is based on detection of the off-diagonal GMI component of amorphous wires [8, 21-23].

**Table 1.** Comparison of features of different sensors.

| Sensor | Head length | Sensitivity / Range | Frequency range | Power consumption |
|---|---|---|---|---|
| Hall sensor | 10-100 μm | 50 μT/ ± 100 mT | DC-1 MHz | 10 mW |
| GMR sensor | 10-100 μm | 1 μT / ± 2 mT | DC-1 MHz | 10 mW |
| Fluxgate | 10-20 mm | 0.1 nT/ ± 300 μT | 1 Hz-5 kHz | 1 W |
| GMI sensor | 0.5 mm | 0.01 nT / ± 3 mT | DC-1 MHz | 10 mW |

|   best |   acceptable |   not acceptable |
|---|---|---|

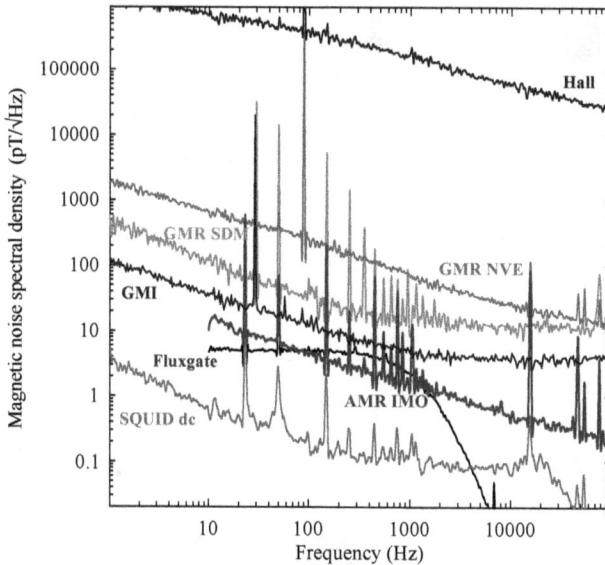

**Figure 28.** Comparison of the magnetic field noise level of some magnetic sensors. Reprinted with permission of authors and the journal from Ref. [60] D. Menard, M. Britel, P. Ciureanu, and A. Yelon, Equivalent Magnetic Noise Limit of Low-Cost GMI Magnetometer, *J. APPL. PHYS.*, V. 84, 5 (1998), 2805-2814. Copyright [1998], AIP Publishing LLC.

The practical laboratory circuit design [8, 21-23] consists of pulse generator, sensor element and output stage. The sharp pulses are produced by passing square wave multi-vibrator pulses through differential circuit. The optimal pulse edge time is reported [21-23] to be about 5 ns, which corresponds to maximum frequency of 50 MHz. This pulse passes through the magnetic wire and the magnetic field dependent signal appears in the pickup coil (Fig. 29).

As mentioned above, the shape of magnetic field dependence of the GMI effect (including off-diagonal components) is intrinsically related with the magnetic anisotropy and peculiar surface domain structure of amorphous wires [20-23]. Magnetic anisotropy of amorphous microwires in the absence of magnetocrystalline anisotropy is determined mostly by the magnetoelastic term [9, 19-23]. On the other hand recently considerable GMI hysteresis has been observed and analyzed in microwires [79]. This GMI hysteresis has been explained through the helical magnetic anisotropy [79].

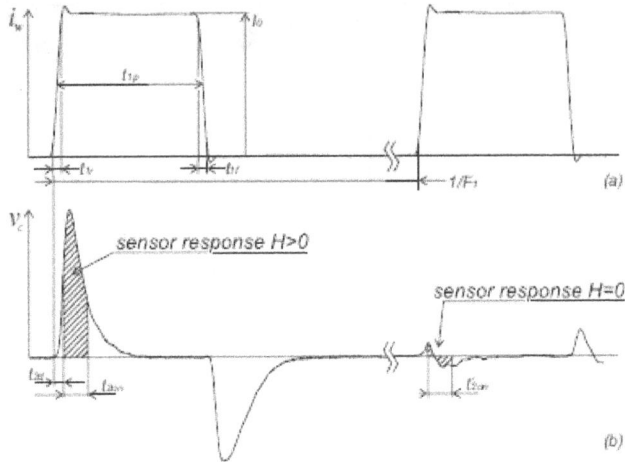

**Figure 29.** Excitation current pulse in the wire (a) and voltage induced in the pickup coil (b). Reprinted with permission from Ref. [76] M. Ipatov, V. Zhukova, J. M. Blanco, J. Gonzalez, and A. Zhukov, Off-diagonal magneto-impedance in amorphous microwires with diameter 6–10 μm and application to linear magnetic sensors, *Phys. Stat. Sol.* (a) 205, No. 8, (2008), 1779–1782. Copyright (2008) WILEY (Fig. 2).

For GMI applications in magnetic field sensors improved linearity of the output signal and low hysteresis are required [12, 90, and 91]. The hysteresis up to 100 A/m or even higher was found in amorphous microwires [79] which considerably limits the sensor's precision. Though the MI effect has been rather extensively studied over the last two decades, the problem of the low-field hysteresis was consider only in a few works [72, 79, 93], where it was shown that the MI hysteresis is related with static circumferential magnetization and the application of circumferential *dc* bias field *HB* is required to suppress this hysteresis.

Recently modelling and even experiments for feasibility studies related with utilization of thin wires with GMI for designing of new generation of magnetic field sensors with improved features [79, 94].

There, the energy minimization of the mono-domain surface layer has been considered for describing the static magnetization reversal process. The equilibrium angle is found to depend on external axial magnetic field $H_E$, circumferential bias field $H_B$, anisotropy field $H_A$ and anisotropy angle $\alpha$. While the last two parameters are constant (although they can be tailored by different annealing procedures or changing the wire geometry [9, 19], $H_E$ and $H_B$ can be used to control the magnetic state and, therefore, the GMI dependence on applied magnetic field. For a wire with helical anisotropy ($\alpha \neq 0$), the impedance, $Z$, becomes sensitive to the circumferential magnetic bias field $H_B$ and exhibits a hysteresis as a function of it.

Thus, increasing the bias field up to 0.5 $H_A$ results in the disappearance of the hysteresis, as it can be seen from the dependence in Fig. 30 for diagonal $Z_{zz}$ and in Fig. 31 off-diagonal impedance $Z_{\phi z}$, components. In the absence of the bias field, both dependencies $Z_{zz}$ and $Z_{\phi z}$ exhibit hysteresis. The application of the bias field $H_B$ makes the dependencies unhysteretic with a high sensitivity $dZ/dH$ slope.

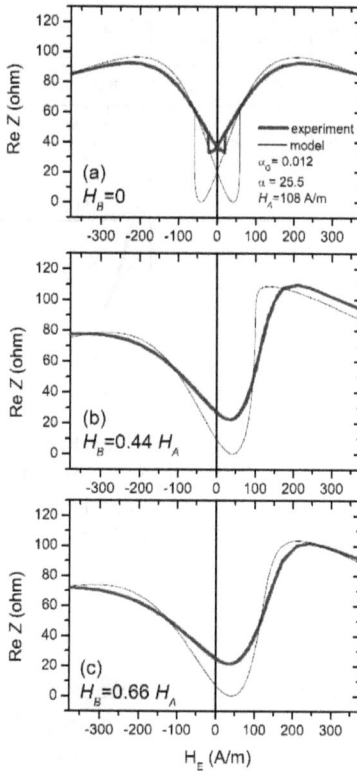

**Figure 30.** Experimental measurement of $Z_{zz}$ and comparison with the model. Reprinted with permission from Ref. [79] M. Ipatov, V. Zhukova, A. Zhukov, J. Gonzalez, and A. Zvezdin, Low-field hysteresis in the magnetoimpedance of amorphous microwires, *Physical Review B*, 81, (2010), 134421, DOI:10.1103/PhysRevB.81.134421. Copyright (2010), The American Physical Society (Fig. 8).

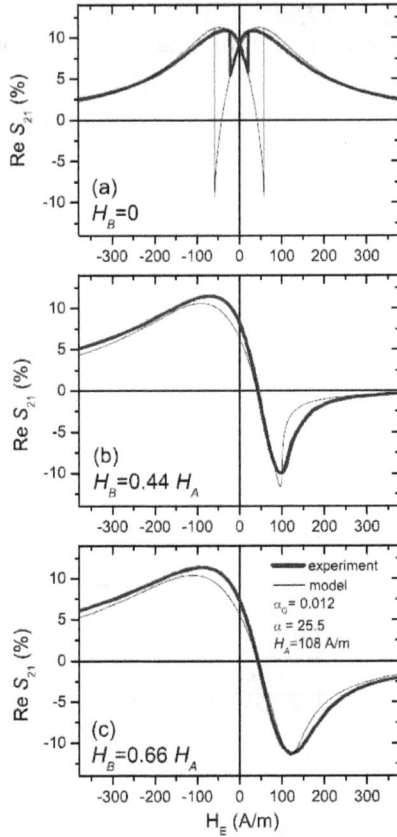

**Figure 31.** Experimental measurement of $Z_{\phi z}$ and comparison with the model. Reprinted with permission from Ref. [79] M. Ipatov, V. Zhukova, A. Zhukov, J. Gonzalez, and A. Zvezdin, Low-field hysteresis in the magnetoimpedance of amorphous microwires, *PHYSICAL REVIEW B*, 81, (2010), 134421, DOI:10.1103/PhysRevB.81.134421. Copyright (2010) The American Physical Society (Fig. 9).

Fig. 32 shows the measured impedance Re $Z$ dependencies on external axial magnetic field $H_E$ with the dc bias current $I_B$ as a parameter. As one can see, at $I=0$, the $Z(H_E)$ curve is symmetric and hysteretic. The hysteresis with the switching field of about 27A/m is observed [see Fig. 31 (c)], which is caused by deviation of the anisotropy easy axis from the transverse direction [79]. To suppress the hysteresis, the application of a bias field $H_B \geq H_A \sin \alpha = 150$ A/m (or corresponding current $I_B = 10$ mA, in this case) is required.

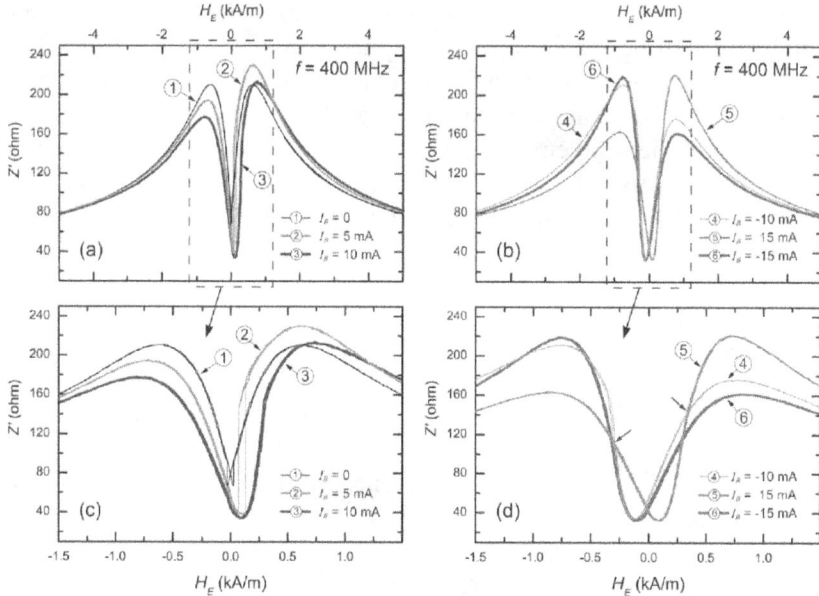

**Figure 32.** Experimental impedance dependencies $Z$ on external axial magnetic field $H_E$ with the dc bias current $I_B$ as a parameter. Graphs (c) and (d) show the enlarged view of the central part. Reprinted with permission from Ref. [94] M. Ipatov, V. Zhukova, J. Gonzalez and A. Zhukov, Symmetry breaking effect of dc bias current on magnetoimpedance in microwire with helical anisotropy: Application to magnetic sensors, *Journal of Applied Physics,* 110, 086105 (2011), Copyright [2011], AIP Publishing LLC. (Fig. 1).

An important effect of the application of a bias field to wire with high helical anisotropy is that it makes the MI curve highly asymmetric. As one can see from Fig. 32, if $I_B$ =5 mA is applied, the MI dependence (curve 2) becomes asymmetric, but the hysteresis is still present. When $I_B$ is increased to 10 mA (curve 3) and 15 mA (curve 5), the hysteresis disappears and the minimum of MI curve shifts to 125 A/m. These curves are mirrored symmetrically with the impedance minimum at $H_E \approx$ -125 A/m when the polarity of $I_B$ is changed (curves 4 and 6).

The observed effect of sensitivity of the wire impedance to the static current flowing through it can be used for different applications. First, we showed the way to decrease GMI hysteresis applying bias current. The other straightforward applications are sensors of the dc current: The current creates the circumferential static magnetic field $H_B$ which rotates the magnetization vector that leads to change of the impedance of the conductor. Similarly, as the effect depends on the anisotropy

65

angle $\alpha$, it can be used in stress or torque sensors - the impedance of the conductor will be sensitive to the applied torsion and tensile stresses. Another possible application is the control of scattering spectra of a microwire-based composite material with dc current. This is an energy saving and easier implementing alternative to the magnetic field control which requires use of bulky coils for generating magnetic fields. Also, as the impedance dependence is hysteretic, it can be used in memory elements.

On the other hand continuous efforts have been performed to improving the sensitivity of the sensor by optimizing the processing parameters and/or the design of the electrical circuit [12, 90-91, 95-101]. In most sensors, the sensing elements can be amorphous wires [12, 90, 91, 95, 101], thin films [96], or ribbons [97]. They can be used for measuring or tracking the presence of both homogeneous and inhomogeneous magnetic fields. A variety of GMI sensors using amorphous wires have been designed and developed by the Aichi Steel Corporation in Japan for a wide range of technological applications [12]. GMI sensors provide several advantages (e.g., low power consumption, small dimension) over conventional magnetic sensors, but their high sensitivity is the most important of these. Thus, use of GMI sensor for detection of magnetic field of the range of pT allowing measuring magnetic fields of brain or blood has been reported [90-91].

The Aichi Steel Corporation has recently designed and produced an ultra small one-package compass device (i.e., the so-called three-axis electronic compass), which allows the sensing of geomagnetic fields with two-axis GMI sensors [101]. One of the previous GMI sensors won a Sensors Expo Silver Award at the recent US Sensors Expo & Conference. For mobile phone applications, the Aichi Steel Corporation has also developed a new device, i.e., the so-called G2 motion sensor, using the GMI effect to detect both geomagnetism and gravity. These G2 motion sensors are being widely used in mobile devices by the Korean mobile phone manufacturer and by the Japanese Vodafone KK Corporation [101-103].

Additionally the magnetometer based on the off-diagonal GMI effect in Co-rich glass-coated microwire has been recently reported [104]. The sensing element in this case is 10 mm long piece of Co-Fe-Ni-B-Si-Mo microwire with a small pick-up coil of 85 turns wounded around the microwire. In this magnetometer the electronics with a feedback circuit has been used to register an electro-motive force proportional to external magnetic field applied along the wire axis. In the absence of the feedback current the magnetometer is capable to measure a narrow range of magnetic fields, $\pm 3.5$ $\mu$T. In the frequency range of 0 - 1 kHz the level of the equivalent magnetic noise was about 10 pT/Hz1/2 at frequency of 300 Hz. The use of the feedback circuit increases the range of the measured magnetic fields up to $\pm 250$ $\mu$T[104]. The magnetometer has a small size, a sensitivity 10 pT/Hz1/2 at frequencies above 300 Hz. The dynamic range of the magnetometer exceeds 120 dB. Both the one – and three – channel versions of the GMI magnetometer are manufactured [104].

## 4.2. Tunable Metamaterials

Recently studies of tunable composite materials containing magnetically soft inclusions gained considerable attention [31, 32, 105]. These materials are interesting for the applications as a multifunctional structural materials and biomedical engineering. On the other hand it was shown, that composites consisting of magnetic metallic wires inside non-magnetic matrix can have both effective permittivity $\varepsilon_{ef}$ and permeability $\mu_{ef}$ at microwave frequencies. A special feature of these composites is that both parameters can demonstrate a strong tunability with respect to varying magnetic structure in wires with such external stimuli as magnetic field, mechanical load and heat. Furthermore, incorporating arrays of magnetic wires in fibre-reinforced polymer composites has also a potential to engineer materials with required structural and electromagnetic functionalities.

Large values of permittivity can be engineered utilizing ferroelectric or conducting elements. The latter could be preferable since very large values of $\varepsilon_{ef}$ are obtained for small volume concentrations and various frequency and spatial dispersions of permittivity are realized. Recently arrays of continuous wires gained much attention as systems with negative real part of the effective permittivity to constitute the materials with left-handed properties by combining the wire arrays and ring resonators [106, 107]. In composites with short-cut wires the length of which is comparable with the wavelength, the effective permittivity may have resonance dispersion in the GHz range due to induced dipole moments of wires resonating at half wavelength condition [107]. This differs greatly from natural dielectrics, where the charge oscillation resonances become important only at optical frequencies.

The magnetic properties of composites materials could be originated by incorporating a ferromagnetic phase. At frequencies of 1-10 GHz, ferromagnetic components in the form of thin-films often provide an optimal response due to their high magnetizations saturation, reduced demagnetizations effects and weak skin effect. Similar performance could be achieved utilizing magnetic wires with special circumferential magnetic anisotropy. It is demonstrated, that the effective permittivity of magnetic wire arrays shows strongly tunable behaviour. Such composites could be of interest for reconfigurable microwave devices as well as for sensory materials [30, 31, 105].

Adjustability of electromagnetic properties is important for many applications, especially in communication, defence and non-destructive testing. This will be highly needed in realization of reconfigurable local network environment, beam steering antennas, and microwave methods of remote sensing and control. A relatively new technology of magnetic wire arrays to manipulate the collective electric response from composite systems has been recently introduced [30, 31, 105].

In thin conducting wires the currents that are responsible for effective permittivity are constrained with the associated resonances determined by the geometrical parameters. The current resonances are damped due to the wire impedance which may increase greatly when the wire magnetization is changed. This is known as giant magnetoimpedance (GMI) effect [14, 15]. In soft magnetic amorphous wires subjected to an external magnetic field GMI is in the range of 100 % even at frequencies of few GHz. Increase in magnetic losses results in increase in the relaxation parameter which determines the frequency dispersion of the effective permittivity. In the case of plasmonic wire arrays, this will result in considerable decrease in the absolute value of the permittivity and will enhance the wave propagation. Similarly, in cut-wire composites, the increase in relaxation broadens the permittivity dispersion which may even show transformation from resonance to relaxation behaviour. Therefore, in composites containing ferromagnetic wires exhibiting GMI effect at GHz frequencies the effective permittivity will depend on the wire magnetic properties via the corresponding dependence of its impedance. Applying a magnetic field $H_{ex}$ larger than the magnetic anisotropy field $H_K$ in wires which is just in the range of few Oersted (fraction of mTesla), rotates the magnetizations towards the axis, and strongly increases its impedance (for frequencies from MHz towards 10 GHz) [105]. Then the permittivity behaviour is damped. Another benefit of using magnetic microwires is that it will be possible to engineer low density materials with relatively high values of the effective magnetic permeability originated from natural magnetic properties of the wires with a circumferential magnetic anisotropy. The magnetic field in the incident wave along the wire will generate substantial magnetic activity as it will be in the orthogonal position with respect to static magnetizations. The demagnetizing effects will not deteriorate the axial permeability as the ac magnetizations could lay in the tangential position to the wire surface. Integrating electric and magnetic wire arrays, composites with relatively large values of both $\varepsilon_{ef}$ and $\mu_{ef}$ could be realized. For example, it will be possible to achieve a negative index of refraction. As comparing to other types of left handed materials, an enhanced performance in terms of tunability, simple internal structure, reduced losses and low cost is anticipated.

Composites containing elongated metallic inclusions can be designed to have specific frequency spectra. In particular, composites with metallic wires can have effective permittivity of plasmonic type (continuous wires), resonant type (wire length is comparable with the wavelength) or relaxation type (wire length is much smaller than the wavelength). Typically, it is considered that the wire radius is small compared to the wavelength. Then, every wire can be described in terms of effective linear current referred to the wire axis, the polarization properties of which determine the effective permittivity. It is further assumed that the wires are ideally conductive; therefore, the field distribution inside them is ignored. This is justified when the skin effect is strong.

In magnetic wires, if the skin effect is essential the loss parameter is enhanced by the wire dynamic permeability. This will make it possible to change the dispersion of the effective permittivity by changing the wire magnetic properties [105, 109]. However, the skin effect should not be too strong when the relaxation is indeed small and the internal properties of constituent wires have little effect on the permittivity spectra. For electrical systems, two main configurations are of interest as shown in Fig. 33: periodic arrays of continuous wires [103, 104] and short-cut wires arranged randomly or periodically [105, 108]. In the first case, the effective permittivity is of a plasmonic type with the plasma frequency determined by the spacing between the wires $b$ and the wire radius $a$. In the second case, the permittivity is of a resonance type in the frequency band within which the half wave length condition is realized: $f_{res}=c/2l\varepsilon_d$ where $l$ is the wire length, $c$ is the velocity of light and $\varepsilon_d$ is the permittivity of matrix. The volume concentration of wires should be below 0.02 %. To design composite with magnetic properties, it is needed to add a magnetic subsystem with much larger concentration (~5-10 %) of magnetic wires which should be placed parallel to the magnetic field of the incident electromagnetic wave and perpendicular to the electrical subsystem.

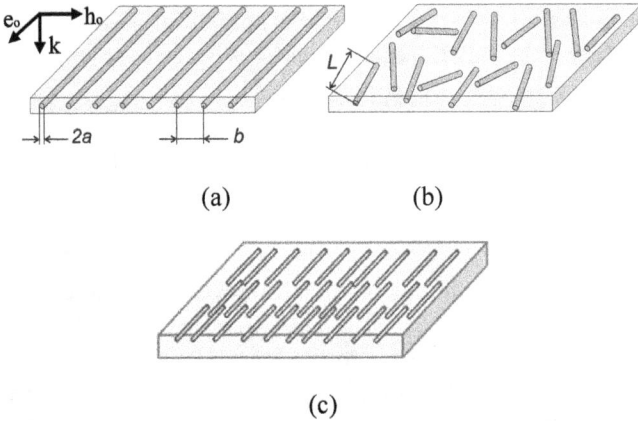

(a)　　　　　(b)

(c)

**Figure 33.** Sketch of wire composites to engineer permittivity spectra. In periodic arrays, the electric field in the incident wave is parallel to the wires. Reprinted with permission from Ref. [105] L. Panina, M. Ipatov, V. Zhukova, J. Gonzalez and A. Zhukov, Tunable composites containing magnetic microwires, Chapter 22, pp. 431-460, Book: Metal, ceramic and polymeric composites for various uses, Edited by John Cuppoletti, ISBN: 978-953-307-353-8 (ISBN 978-953-307-1098-3), InTech - Open Access Publisher (www.intechweb.org), Janeza Trdine, 9, 51000 Rijeka, Croatia (Fig. 1).

We demonstrated that the dispersion properties of permittivity in magnetic wire media depend on the wire internal magnetic structure following the magnetic behavior of the wire impedance. Combining the dispersion properties of wire media and GMI effect it is possible to actively tune the permittivity spectra of arrays of magnetic microwires by application of a small magnetic field and a stress which is now demonstrated in a number of experimental studies (see below). However, to realize large and sensitive tuning requires the existence of very special magnetic structures in wires as that with a circular anisotropy. The methods of tailoring magnetic structure in amorphous wires have been discussed above.

It is seen that in the presence of the external bias field the dispersion region broadens since the losses are increased in high impedance state of the wires (Fig. 34).

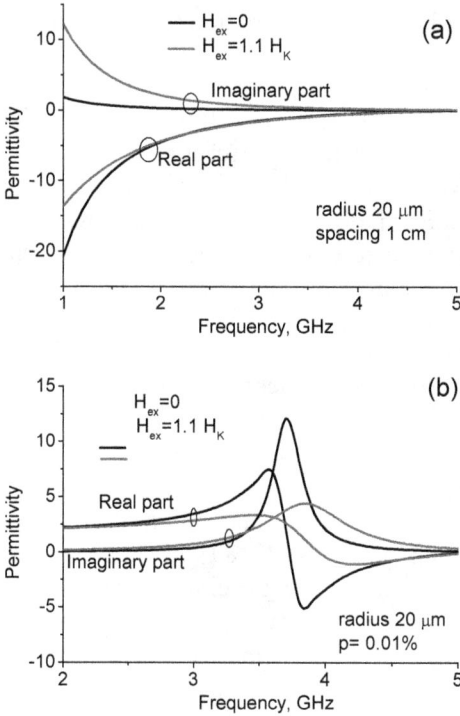

**Figure 34.** Effective permittivity spectra in composites depicted in Fig. 33 with the external field as a parameter. Modeling is performed for wires with a circumferential anisotropy (anisotropy field $H_k$=500A/m). The other parameters are: resistivity 130 $\mu\Omega$·cm, magnetization 0.05T, wire radius 20 $\mu$m. For (a), $b$=1 cm. For (b), $l$=4 cm, $p$=0.01 %. Reprinted with permission from Ref. [105] L. Panina, M. Ipatov, V. Zhukova , J. Gonzalez and A. Zhukov, Tunable composites containing magnetic microwires, Chapter 22, pp. 431-460, Book: Metal, ceramic and polymeric composites for various uses, Edited by John Cuppoletti, ISBN: 978-953-307-353-8 (ISBN 978-953-307-1098-3), InTech - Open Access Publisher (www.intechweb.org), Janeza Trdine, 9, 51000 Rijeka, Croatia (Fig. 4).

Since its discovery in 1994 [14, 15] the GMI effect has become a topic of great interest in the field of applied magnetism owing to the large sensitivity of the total impedance to the applied DC field at low field magnitudes and high frequencies.

As mentioned above, stress-impedance (SI) and torsion impedance (TI) effects showing high sensitivity of the impedance to the applied stress with a strain gauge factor of 2000-4000 have been found in amorphous wires [8, 9]. Main applications of GMI effect are related with the detection of the magnetic fields, small weights, vibrations, acceleration, and recently, in microwave sensory and tunable composites.

Enhanced GMI ratios in amorphous wires were explained by a specific domain structure existing in negative magnetostrictive amorphous wires [14, 15]. Negative magnetostriction coupled with tensile stress creates alternative left and right handed circular domains in the outer sheath of wires. Such domain configuration was observed in wires produced by various methods including in-rotating water developed by Unitika LTD and in glass-coated microwires [8, 9].

As mentioned above, one of the most important effects occurring due to stress annealing is enhancement of stress sensitivity of magnetic properties and stress- impedance [8, 9, 10-32, 83, 84, 105]. In particularly, it was demonstrated, that stress annealing, performed at certain annealing conditions results in induction of stress-sensitive transverse magnetic anisotropy and observation of significant (up to 60 %) stress-impedance effect [83, 84], as shown in above Fig. 19. This result is of special interest for developing stress-sensitive composites with the use of magnetic microwires. The origin of this creep annealing induced anisotropy has been attributed to redistribution of the residual stresses during the stress annealing which results in drastic decrease in the longitudinal stress component and even in the appearance of the compressive longitudinal stresses (so-called "back stresses").

Typically, the Curie temperature of Fe and Co-rich amorphous microwires is about 300-400 °C. Additions of Ni and Cr in the alloys make it possible to substantially decrease the Curie temperature [105, 1110, 111]. For example, the Curie temperature between 75 and 90 °C was reported for microwires with composition $Co_{60.51}Fe_{3.99}Cr_{12.13}B_{13.53}Si_{9.84}$ and $Co_{23.67}Fe_{7.14}Ni_{43.08}B_{13.85}Si_{12.26}$. This suggest a potential to develop magnetically soft microwires showing large temperature dependence of magnetization, anisotropy, magnetic permeability etc. (Fig. 35) and, hence, GMI effect. Then, the microwires would be suitable for remote temperature detection in the range of moderate temperatures from room temperature to about 400 °C.

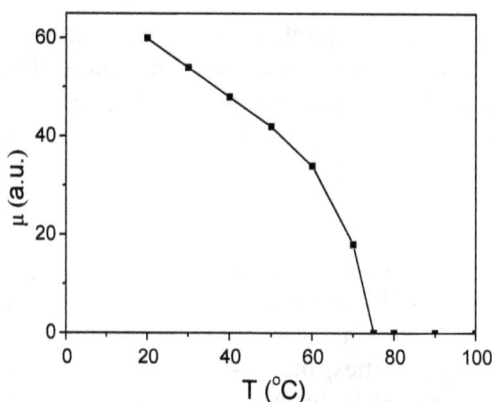

**Figure 35.** Temperature dependence of permeability (given in arbitrary units) measured in $Co_{60.51}Fe_{3.99}Cr_{12.13}B_{13.53}Si_{9.84}$ microwire. Reprinted with permission from Ref. [111] V. Zhukova, J. M. Blanco, M. Ipatov, A. Zhukov, C. Garcia, J. Gonzalez, R. Varga, A. Torcunov, Development of thin microwires with low Curie temperature for temperature sensors applications, *Sensors and Actuators B*, 126 (2007) 318–323 (Fig. 4) .

The developed magnetic field, stress, and temperature - sensitive microwires have been proposed for completely new range of applications as constituent elements of wire-composites for tunable microwave systems, and non-destructive remote control of stress, strain and temperature [30-32, 73, 105]. The wires can be regarded as embedded sensors and their impedance sensitive to the wire magnetic structure will be responsible for producing a controlled microwave dielectric response.

The material parameters in microwaves frequencies usually are found from the measurement of the reflection and/or transmission coefficients from which the complex permittivity and permeability are calculated. The measurement methods can be divided in two categories: (i) transmission line methods (coaxial lines probes, rectangular waveguides, and cavity resonators) and (ii) antenna techniques in free space. The methods in the first category require cutting a piece of a sample to be placed inside the transmission line or cavity making a close contact with the probe. The transmission line methods work best for homogeneous materials that can be precisely machined to fit inside the sample holder.

At microwaves, the measurement of the effective permittivity of composite materials with the inhomogeneity scale comparable with the wavelength requires large sample dimensions. In this case, the method of spot localized measurement area, such as conventional coaxial line and waveguide methods cannot be used. A free-space method is more appropriate. Generally, it is used to characterize large flat solid materials, although granular and powdered materials can also be measured in a fixture.

Free-space techniques for material property measurements have several advantages [105]. Firstly, materials such as ceramics and composites are inhomogeneous due to variations in manufacturing processes. Because of inhomogeneities, the unwanted higher-order modes can be excited at an air-dielectric interface in hollow metallic waveguides, while this problem does not exist in free-space measurement. Secondly, the measurements using free-space techniques are non-destructive and contactless, so free-space methods can be used to measure samples under special conditions, such as high temperature. Thirdly, in hollow metallic waveguide methods, it is necessary to machine the sample so as to fit the waveguide cross section with negligible air gaps. This requirement limits the accuracy of measurements for materials that cannot be machined precisely; in free-space method, this problem does not exist. Finally, waveguides have a rather narrow operating frequency range. Therefore, to characterized material in a wide frequency range, a number of waveguides is required. Moreover, every waveguide, having different cross-section, will required preparation of separate sample.

Further we consider the free space method for measurement of the electromagnetic parameter of the composites as the most suitable for both laboratory investigation and *in situ* non-destructive testing and remote structural health monitoring. In free space method, materials are placed between antennas for a non-contacting measurement allowing much flexibility in studying materials under different conditions such as high temperatures and hostile environments. A key component of any free space system is antenna that is a transition element between transmission line and free space radiating and/or receiving the electromagnetic energy into/from free space.

The experimental setup for the reflection/transmission microwave free-space measurements basically consists of vector network analyzer (VNA), a pair of broadband horn antennas and an anechoic chamber as shown in Fig. 35. A composite sample is placed in the middle of the chamber with the wire orientation along the electric-field of the incident electromagnetic wave. The desired frequency range, in which the scattering parameters will be investigated, determines the requirements to the operating frequency of the VNA, antennas and to the chamber size (distance between antennas and sample). The lens can be applied to focus the radiation pattern and minimize the effect of sample boundaries and measuring environment. It is essential to place the sample outside the reactive near-field region where the wave is not polarized and the electromagnetic interaction between the sample and the antenna can arise. The reactive near-field terminates at the distance of the order of wavelength $\lambda$ from antennas.

Free space method imposes a limit on the minimal sample size. If the sample size is much smaller than the wavelength, the response of the sample to electromagnetic waves will be similar to those of a particle object. To achieve convincing results, the size of the sample should be larger than the wavelength of the electromagnetic wave. To further minimize the effects of the scatterings from the sample boundary, the sample size should be at least twice larger than the wavelength [105]. Therefore if the lowest measurement frequency is 1 GHz, then the sample size should be $2\lambda = 60$ cm.

The free space technique requires precise calibration. The Thru-Rreflect-Line (TRL) and Thru-Reflect-Match (TRM) calibration techniques, that were commonly used until recently, are being widely replaced with the Gated Reflect Line (GRL) calibration [105]. The GRL calibration, based on the time domain gating, allows enhancing the calibration accuracy and elimination of the need for expensive spot focusing antennas and micro positioning fixturing. Time-domain technique is important not only for calibration of the free-space measurement path but also during the measurements as it makes it possible to effectively eliminate the effects of multipath reflections to which the measurements in free-space are subjected. The main source of the reflection is the inevitable mismatch between the antenna and free space. The other error sources such as reflection from chamber's walls and noise could be also essential. The time domain procedure "gates out" these error terms and also reduces the requirements for the quality of the anechoic chamber or even allowing conducting the measurements without the chamber.

The free space setup, shown in Fig. 36 is applied for the wire composite characterization. It consists of Agilent 0.01 – 20 GHz two port VNA with time domain option, two broadband horn antennas with the operating range 0.9 – 17.0 GHz and the anechoic chamber with dimensions $80 \times 80 \times 80$ cm$^3$ covered inside with a microwave absorber. The composite sheets are placed at a distance of 40 cm from each horn antenna, appearing in the radiating near-field region in the whole range of operating frequencies. 85071E Material Measurement Software (Agilent) and "Reflection/Transmission Epsilon Fast Model" can be used for calculating the complex permittivity of the composites from the experimental S-parameters.

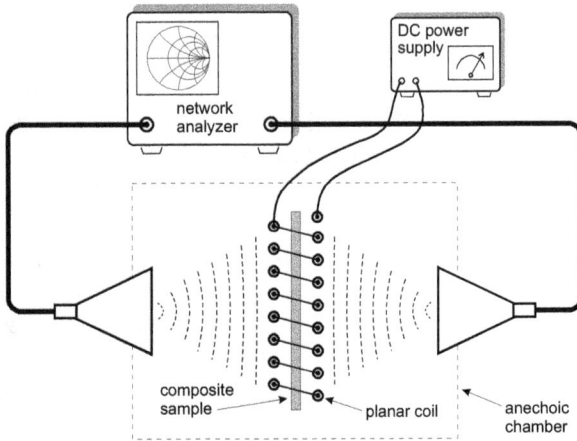

**Figure 36.** Sketch of the free space microwave measurement setup. Reprinted with permission from Ref. [105] L. Panina, M. Ipatov, V. Zhukova, J. Gonzalez and A. Zhukov, Tunable composites containing magnetic microwires, Chapter 22, pp.431-460, Book: Metal, ceramic and polymeric composites for various uses, Edited by John Cuppoletti, ISBN: 978-953-307-353-8 (ISBN 978-953-307-1098-3), InTech - Open Access Publisher (www.intechweb.org), Janeza Trdine, 9, 51000 Rijeka, Croatia (Fig. 14).

To study the influence of magnetic field on the dispersion characteristics of the composite samples a special planar magnetic coil was constructed as described in [73]. A thin planar composite sample is placed inside the magnetic coil so that the microwires were along the direction of magnetic field. The coil having the field coefficient 90A/m/A creates a homogeneous along the sample surface DC magnetic field. A 35 Ampere Agilent 6674A DC power supply is used to feed the coil and permits to reach the magnetic field as high as 3000 A/m with resolution below 1 A/m. The coil turns are set perpendicular to the electrical field so there is no effect of the coil on scattering.

In a simple way, the wire composites can be prepared by gluing them on paper to form wire-lattices of needed dimensions. Firstly, 1D wire-lattices of plasmonic type with continuous wires (as shown in Fig. 33a) are arranged. Composites with short- wire pieces forming ordered electrical dipoles (seen in Fig. 33c) could be obtained by cutting the continuous wire arrays in stripes of different size to be able to change the dipole resonance frequencies. The wires used for these experiments were glass coated microwires of the composition $Co_{66}Fe_{3.5}B_{16}Si_{11}Cr_{3.5}$, with magnetic core diameter of 20 microns. These wires have small but negative magnetostriction and show GMI in the range of 100 % at few GHz [105].

Fig. 37 shows the spectra of the reflection $R$ and transmission $T$ for composites with continuous wires having spacing $b=1$ cm with the magnetic field $H_{ex}$ as a parameter. The relative change of $T$ (Figure of Merit $FOM=(T(0)-T(H_{max})/T(0))$ is about 25% at 1.8 GHz while the phase of transmission shifts about 40 degrees with the field change from 0 to 500 A/m. Fig. 38 shows the absorption parameter $A=1-R^2-T^2$, which changes by 4 times from 10 % to 40 %.

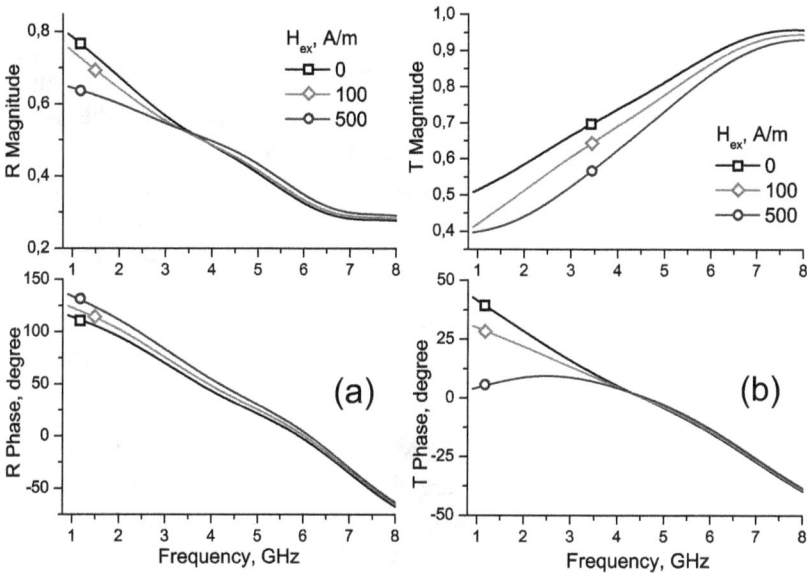

**Figure 37.** Spectra of $R$ and $T$ for composites with continuous wire arrays with $H_{ex}$ as a parameter. Reprinted with permission from Ref. [105] L. Panina, M. Ipatov, V. Zhukova, J. Gonzalez and A. Zhukov, Tunable composites containing magnetic microwires, Chapter 22, pp.431-460, Book: Metal, ceramic and polymeric composites for various uses, Edited by John Cuppoletti, ISBN: 978-953-307-353-8 (ISBN 978-953-307-1098-3), InTech - Open Access Publisher (www.intechweb.org), Janeza Trdine, 9, 51000 Rijeka, Croatia (Fig. 14).

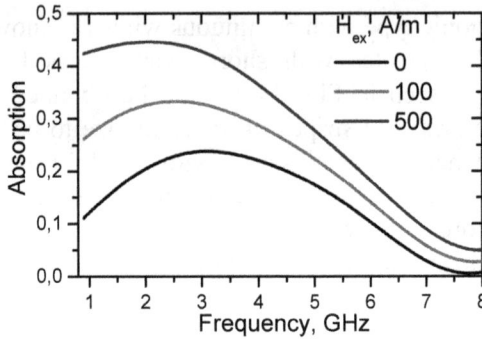

**Figure 38.** Absorption spectra for composites with continuous wire arrays $H_{ex}$ as a parameter. Reprinted with permission from Ref. [105] L. Panina, M. Ipatov, V. Zhukova, J. Gonzalez and A. Zhukov, Tunable composites containing magnetic microwires, Chapter 22, pp.431-460, Book: Metal, ceramic and polymeric composites for various uses, Edited by John Cuppoletti, ISBN: 978-953-307-353-8 (ISBN 978-953-307-1098-3), InTech - Open Access Publisher (www.intechweb.org), Janeza Trdine, 9, 51000 Rijeka, Croatia (Fig. 15).

The permittivity spectra shown in Fig. 39 were calculated from $R$ and $T$ coefficients. The effective thickness was taken equal to the lattice period $b$ although the real composites thickness is much smaller and is defined by the host matrix thickness. The real part of $\varepsilon_{ef}$ has a negative value below the plasma frequency which is equal to 4.7 GHz. Application of the field decreases the value of the real part (compare with theoretical plots shown in Fig. 4) as the wire impedance increases and so does the relaxation. The imaginary part of the permittivity, that directly demonstrates the losses or absorption, increases with the field. This results in reduced transmitted and reflected signals.

The $R$ and $T$ spectra for the cut-wire composites with different wire length $l$ of 40, 20 and 10 mm are shown in Fig. 40. The transmission spectra have a deep minimum near a resonance demonstrating stop-filter behaviour. The magnitude of this minimum depends strongly on the field for longer wires with lower resonance frequency. For 40 mm long-wire composites, FOM = 102 % at 3.57 GHz and for 20 mm long wires FOM=57 % at 6.18 GHz. For shorter wires with the dispersion region at a higher frequency band the field dependence is not noticeable since the wire ac permeability is nearly unity and the impedance becomes insensitive to the magnetic properties, as explained above. The phase of transmission component exhibits a phase reversal and negative grope delay in the stop band region.

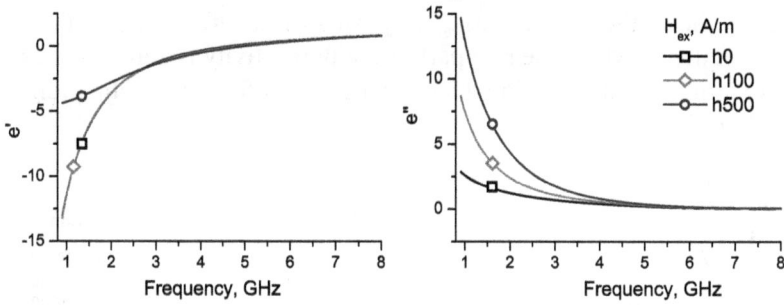

**Figure 39.** Effective permittivity spectra for composites with continuous wire arrays with $H_{ex}$ as a parameter. Reprinted with permission from Ref. [105] L. Panina, M. Ipatov, V. Zhukova, J. Gonzalez and A. Zhukov, Tunable composites containing magnetic microwires, Chapter 22, pp.431-460, Book: Metal, ceramic and polymeric composites for various uses, Edited by John Cuppoletti, ISBN: 978-953-307-353-8 (ISBN 978-953-307-1098-3), InTech - Open Access Publisher (www.intechweb.org), Janeza Trdine, 9, 51000 Rijeka, Croatia (Fig. 16).

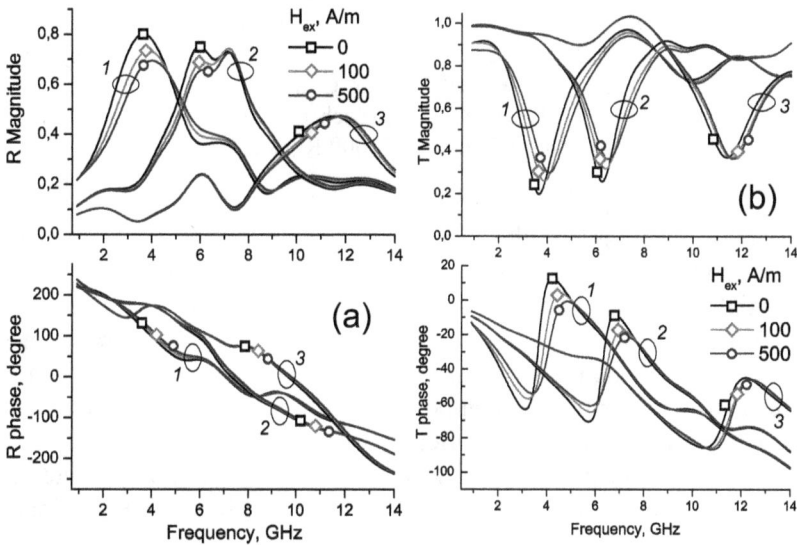

**Figure 40.** Spectra of $R$ and $T$ of composites with cut wires of length 40 (1), 20 (2) and 10 (3) mm with the field as a parameter. Reprinted with permission from Ref. [105] L. Panina, M. Ipatov, V. Zhukova, J. Gonzalez and A. Zhukov, Tunable composites containing magnetic microwires, Chapter 22, pp.431-460, Book: Metal, ceramic and polymeric composites for various uses, Edited by John Cuppoletti, ISBN: 978-953-307-353-8 (ISBN 978-953-307-1098-3), InTech - Open Access Publisher (www.intechweb.org), Janeza Trdine, 9, 51000 Rijeka, Croatia (Fig. 17).

Fig. 41 shows the real and imaginary parts of the effective permittivity. The frequency where the real part of the permittivity is zero depends on the external magnetic field which can be useful for constructing tunable *Epsilon-Near-Zero* materials.

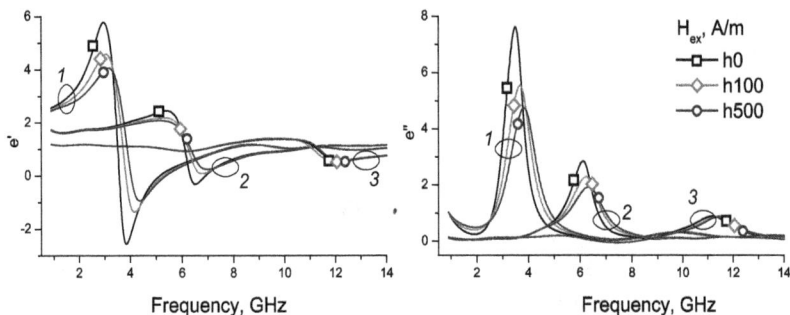

**Figure 41.** Effective permittivity spectra for composites with cut wires of length 40 (1), 20 (2) and 10 (3) mm with the field as a parameter. Reprinted with permission from Ref. [105] L. Panina, M. Ipatov, V. Zhukova, J. Gonzalez and A. Zhukov, Tunable composites containing magnetic microwires, Chapter 22, pp.431-460, Book: Metal, ceramic and polymeric composites for various uses, Edited by John Cuppoletti, ISBN: 978-953-307-353-8 (ISBN 978-953-307-1098-3), InTech - Open Access Publisher (www.intechweb.org), Janeza Trdine, 9, 51000 Rijeka, Croatia (Fig. 17).

Ferromagnetic microwire based composites with tunable electromagnetic characteristics represents a new technology with potentially wide applications. The potential applications for composite based on microwires with large and sensitive magneto-impedance effect can be divided in two categories. In the first one, the MI effect in wires is used to control the composite's electromagnetic characteristics. The application of a magnetic field or other stimuli will cause change in reflection, transmission and absorption in the composite material. For example, an "active microwave window" can be realized, the state of which can be changed from transparent (open) to opaque (close) for the microwaves. Other applications are transmission signal modulation, deferent frequency selective surfaces and reconfigurable absorbers.

The second category includes different sensing applications that use a high sensitivity of the wire impedance to external stimuli. They include non-destructive testing and structure health monitoring for detection of invisible structural damages and defects, monitoring stress concentrations and temperature distribution. The measurement can be conducted in the waveguide and in the free-space. The later being a non-contact remote method is of special interest as it allows *in-situ* health monitoring of objects such as infrastructure (bridges, buildings, etc.), pipeline and pressure vessels.

As the permittivity $\varepsilon_{ef}$ depends on the wire surface impedance, which in turn is a function of the magnetic permeability of the wire, then any physical phenomenon that affects the permeability will affect $\varepsilon_{ef}$. Applying a stress or a torque to amorphous wires causes the change in the wire transverse magnetizations and very large variations in impedance (SI) [83, 84]. At GHz frequencies, however, sensitive SI effect requires a special magnetic anisotropy [73]. For example, in the case of negative magnetostrictive wires the anisotropy should be nearly axial (customarily, it is circumferential). Only in this case the applied tensile stress may produce large effect on impedance through the change in static magnetizations direction. The SI effect is promising for constructing composites with stress-sensitive dispersion of $\varepsilon_{ef}$. The stress applied to the composite matrix will be transmitted to each wire inclusion through matrix strain. In [73] the stress sensitivity of composite media containing amorphous microwires was demonstrated experimentally.

The magnetic structure and GMI of the ferromagnetic wires can be made highly temperature dependent for moderate temperature regions (50-200 °C) that gives a possibility to construct the composites with thermally tunable microwave response and can find applications in remote temperature monitoring by free-space method, e.g., for composite curing control.

The other approach to realize temperature sensitive composites is based on the ferromagnetic-paramagnetic transition at Curie temperature, $Tc$. This transition is characterized by the drastic change of properties such as magnetization, magnetic susceptibility, anisotropy etc. It could be expected that GMI ratio will also change near $Tc$, constituting basis for remote temperature detection. The addition of Ni and Cr in Co/Fe amorphous alloy systems results in a decrease in $Tc$ down to the room temperature [110, 111]. In this way a wide variety of temperature-sensitive composites based on microwires with the $Tc$ ranging between the room temperature and 400 °C can be realized. At approaching $Tc$, the magnetizations saturation $M_s$ scales with $(1-T/Tc)^{\beta}$ and the magnetostriction scales as $M^{\tilde{z}}$. It will result in increase in the initial rotational permeability proportional to the ratio of the magnetizations and magnetostrictive anisotropy field. However, high frequency properties will deteriorate. Then, the high frequency impedance is expected to decrease near the Curie temperature.

In both proposed approaches, the decrease in the wire impedance as the temperature increases produces substantial changes in the frequency dispersion of the effective permittivity and scattering parameters from such composites. For example, in the case of cut wire composites, the resonance type dispersion of the permittivity and band gap propagation regimes will become more pronounce with increasing temperature. These investigations require modification of the microwave setup such as a construction of special thermo chamber which is quite realistic to be realized.

# 5. Fast Domain Wall Dynamics in thin Wires

## 5.1. Magnetic Bistability

One of the main technological interests for utilization of amorphous microwires is related with Large and single Barkhausen Jump (LBJ) observed above some value of applied magnetic field, called switching field between two stable remanent states. Such particular magnetization process has been observed previously also in iron whiskers [112], amorphous Co-rich ribbons after special heat treatment [113-115] and different kinds of amorphous wires [25, 26, 39, 105, 106, 116-120]. Abrupt magnetization jump and sharp electrical pulses related with the magnetization switching during LBJ have been proposed in different kinds of magnetic sensors, such as magnetic markers and magnetoelastic sensors [25, 26, 39, 114].

It is worth mentioning that appearance of Large and single Barkhausen jump takes place under magnetic field above some critical value (denominated as switching field) and also if the sample length is above some critical value denominated also as critical length. As described above, switching field depends on magnetoelastic energy determined by the strength of the internal stresses, applied stresses and magnetostriction constant determined by the chemical composition of ferromagnetic metallic nucleus [9, 29].

Regarding the critical length, detailed studies of the ferromagnetic wire diameter on magnetization profile and size of the edge closure domains have been performed in [29]. Particularly, critical length, $l_c$, for magnetic bistability in conventional Fe-rich samples (120 μm in diameter) is about 7 cm. This critical length depends on saturation magnetization, magnetoelastic energy, domain structure, magnetostatic energy [29]. Thus, in Co-rich conventional amorphous wires (120 μm in diameter) such critical length is about 4 cm [117-120]. Below such critical length hysteresis loop loses its squared shape.

The magnetostatic energy depends on the demagnetizing field, $H_d$, expressed as:

$$H_d = NM_s, \qquad (10)$$

where N is the demagnetizing factor given for the case of long cylinder with length, $l$, and diameter, $D$, as:

$$N = 4\pi(\ln(2l/D) - 1](D/l)^2) \qquad (11)$$

On the other hand the shape of the hysteresis loop of wires with a given length depends on the applied stress. As an example, Fig. 42 shows the effect of the applied stress on the hysteresis loop of the 8 cm long sample measured by the long secondary coil. It is worth to mention that the initially rectangular shape of the hysteresis loop is lost under the certain applied stress (around 70 MPa) but finally it recovers under stronger applied stress of around 320 MPa.

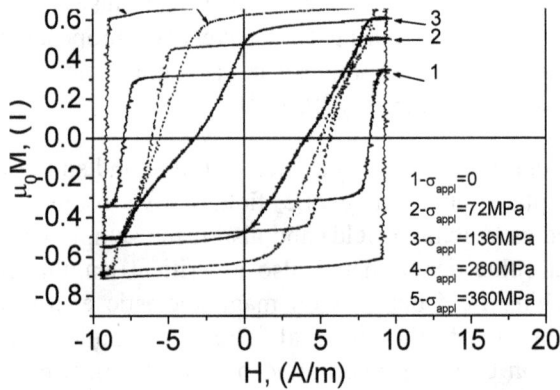

**Figure 42.** Stress dependence of the hysteresis loops of 8 cm long $Fe_{77.5}B_{15}Si_{7.5}$ amorphous wire with 125μm diameter, measured by a long pick-up coil. Reprinted with permission from Ref. [118] V. Zhukova, A. Zhukov, J. M. Blanco, J. González, C. Gómez–Polo and M. Vázquez, Effect of applied stress on magnetization profile of Fe-Si-B amorphous wire, *J. Appl. Phys.*, 93 (2003), 7208-7210. Copyright (2003) AIP Publishing LLC. (Fig.1).

Similar experiments has been done by using short movable pick-up coil which permitted to measure local hysteresis loops as well as to obtain the magnetization profile and its dependence on applied stress. Such dependence measured in the center of the 8 cm long wire, i.e. at the position of the movable pick-up coil of $L=4$ cm and at $L=2$ cm are presented in Fig. 43 (a, b) respectively. At $L=4$ cm the hysteresis loop measured for $\sigma_{appl} < 60$ MPa and $\sigma_{appl} > 320$ MPa exhibit squared shape, while at $L=2$ cm all hysteresis loop are not perfectly rectangular.

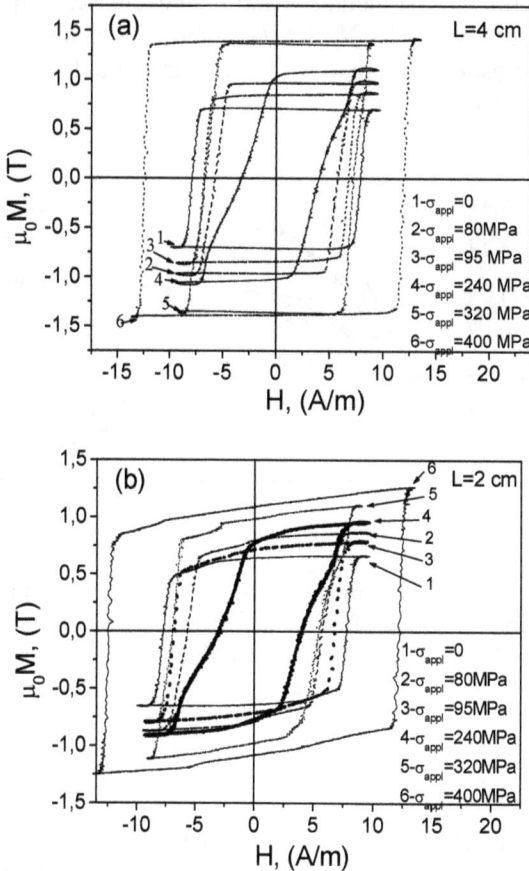

**Figure 43.** Effect of stress applied on hysteresis loops of 8 cm long $Fe_{77.5}B_{15}Si_{7.5}$ amorphous wire, measured by a short movable coil placed at $L=4$ cm (a) and 2 cm (b). Reprinted with permission from Ref. [118] V. Zhukova, A. Zhukov, J.M. Blanco, J. González, C. Gómez–Polo and M. Vázquez, Effect of applied stress on magnetization profile of Fe-Si-B amorphous wire, *J. Appl. Phys.*, 93 (2003), 7208-7210 Copyright (2003) AIP Publishing LLC. (Fig. 2).

The remanence, $\mu_0 M_r$, profile has been acquired from the measurements of local hysteresis loops. Such profiles measured for the

8 cm long wire is shown in Fig. 44. The remanence profile has roughly symmetric character. The decrease of the remanent magnetization close to the wire ends (variation of $\mu_0 M_r$ with the position of the coil, $L$) as well as the variation of the penetration depth, $l_d$, of the closure domains with $\sigma_{appl}$ are observed in Fig. 44. As can be assumed from Fig. 44, the penetration depth, $l_d$ first increases and then decreases with $\sigma_{appl}$. Below certain sample length, $L \approx l_c \approx 2\ l_d$, magnetic bistability disappears because the closure domains penetrate far enough inside the inner axially magnetized core and destroy it.

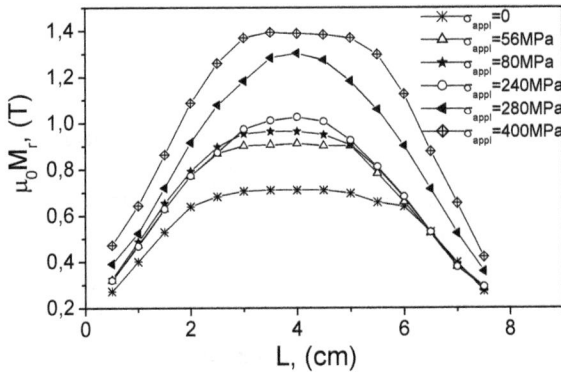

**Figure 44.** Effect of stress on magnetization profiles measured in $Fe_{77.5}B_{15}Si_{7.5}$ amorphous wire measured for sample length $l=8$ cm. Reprinted with permission from Ref. [118] V. Zhukova, A. Zhukov, J.M. Blanco, J. González, C. Gómez–Polo and M. Vázquez, Effect of applied stress on magnetization profile of Fe-Si-B amorphous wire, *J. Appl. Phys.*, 93, (2003), 7208-7210. Copyright (2003) AIP Publishing LLC. (Fig. 5).

Consequently the critical length first increases with $\sigma_{appl}$, but then again decreases and finally magnetic bistability recovers under strong enough applied tensile stress.

General increase of $\mu_0 M_r$ with $\sigma_{appl}$ should be attributed to the increasing of the volume of axially magnetized area under the effect of applied stresses due to the positive magnetostriction constant. On the other hand in the range 60 MPa $\leq \sigma_{appl} \leq 300$ MPa magnetic bistability disappears and the local hysteresis loops do not show rectangular character even in the central region of 8 cm long wire (see Fig. 42). Such behavior is related with the demagnetizing factor of the internal axially magnetized core: the diameter of such single domain core increases which results in increasing of the demagnetizing factor given by eq. (20). This affects the penetration depth of the closure domains, $l_d$, and in this way the magnetic bistability is destroyed even in the central zone of wire.

Further increase of applied stresses results in further increasing of the remanent magnetization and also in drastic increasing of the coercivity and recovering of magnetic bistability. We assume that the increase of the external tensile stress results in the increase of the domain wall energy,

$$\gamma = 2(AK)^{1/2}, \tag{12}$$

where A is the exchange energy and K is the magnetic anisotropy constant, which in the case of amorphous state depends mainly on the magneto-elastic component, $K_{me}$, given by

$$K_{me} \approx 3/2 \, \lambda_s \sigma, \tag{13}$$

where $\sigma = \sigma_i + \sigma_a$ is the total stress, $\sigma_i$ is the internal stresses, $\sigma_a$ is the applied stresses and $\lambda_\sigma$ is the magnetostriction constant.

Applying external stress the magnetoelastic energy increases and, consequently, the domain wall energy, $\gamma$. At such conditions there is a competition between magnetostatic and magnetoelastic energies. Such increasing of the magnetoelastic energy gives rise to the decreasing of the size (and consequently of the penetration depth) of closure domains and therefore results in new decreasing of the critical length and recovering of the magnetic bistability.

Phenomenon of magnetic bistability observed in different families of amorphous wires is certainly quite interesting for the magnetic sensors applications. But, enhanced critical length, $l_c$, (of the order of few cm) observed in conventional amorphous wires limited these applications. On the other hand, reduction of the metallic nucleus diameter in the case of glass-coated microwires (almost one order lower) results in drastic reduction of the critical length, making them quite attractive for micro-sensor applications. Thus, magnetic bistability for the sample length L= 2 mm has been observed for Fe-rich microwire with metallic nucleus diameter, d, about 10 μm [29]. Remanent magnetization profile measured in Fe-rich microwires exhibit deviations just near the sample ends (see comparison of the magnetization profiles in Fig. 45). Strong internal stresses (mostly of axial origin), appearing during the fabrication of glass coated microwires, can also contribute to the reduction of the critical length for magnetic bistability.

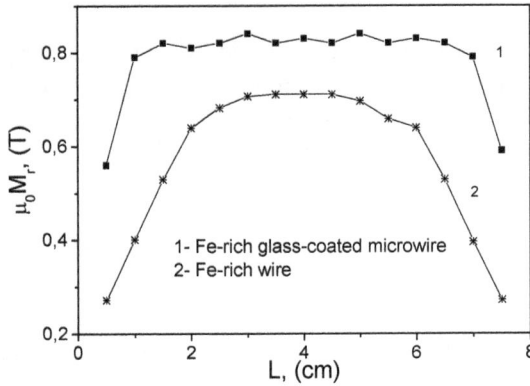

**Figure 45.** Comparison of the magnetization profiles measured in Fe-rich wire with 125μm diameter and Fe-rich microwire. Reprinted with permission from [29] A. P. Zhukov, M. Vazquez, J. Velazquez, H. Chiriac, V. Larin, The remagnetization process in thin and ultra-thin Fe-rich amorphous wires, *J. Magn. Magn. Mat.*, 151,(1995), 132-138. Copyright (1995) with permission from Elsevier (Fig. 4).

One of the extraordinary properties exhibited by amorphous microwires with positive magnetostriction constant is observation of single and large Barkhausen jump, allowing studying details of remagnetization process by relatively simple methods. Nevertheless, the use of these materials is limited, due to intrinsic switching field fluctuations observed in all these material [112-117, 121, 122].

First phenomenological description of switching field fluctuations has been first developed for amorphous ribbons [113-115] and later extended to glass-coated microwires [117, 121, 122] Such switching field fluctuation results in spontaneous experimental spread in the value of the switching field (see Fig. 46) taking such value in different re-magnetizing cycles [113-115].

These switching field fluctuations can deteriorate the performance of the magnetic marker based on a number of encoding combinations because of the overlapping of the switching field values from different microwires and of the magnetoelastic sensor making its work unreliable.

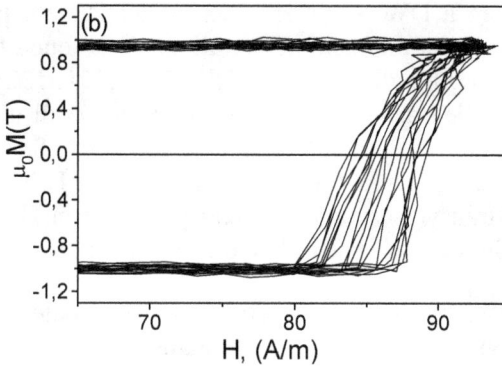

**Figure 46.** Hysteresis loop of the $Fe_{65}Si_{15}B_{15}C_5$ microwire exhibiting switching field fluctuations. Reprinted with permission from [117], V Zhukova, A. Zhukov, J. M. Blanco, J. Gonzalez and B. K. Ponomarev, Switching field fluctuations in a glass coated Fe-rich amorphous microwire, *J. Magn. Magn. Mat.*, 249/1-2, (2002), 131-135, Copyright (2002) with permission from Elsevier (Fig. 1).

## 5.2. Effect of Magnetoelastic Anisotropy

As pointed above, one of the characteristic features of the magnetic bistability is the appearance of rectangular hysteresis loop related with single and large Barkhausen jump appeared at low applied magnetic field. Such magnetic bistable behaviour has been related to the presence of a single Large Barkhausen Jump, which was interpreted as the magnetization reversal in a large single domain [7-9, 18, 26]. The rectangular hysteresis loop could be interpreted in terms of nucleation or depinning of the reversed domains inside the internal single domain and the consequent domain wall propagation [26]. Perfectly rectangular shape of the hysteresis loop has been related with a very high velocity of such domain wall propagation.

Recently great attention has been paid to studies of domain wall (DW) propagation in different wires families [123-130]. Recent growing interest on DW propagation is related with proposals for prospective logic [123] and memory devices [124]. In these devices, information is encoded in the magnetic states of domains in lithographically patterned nanowires. DW motion along the wires allows for the access and manipulation of the stored information. The speed at which a DW can travel in a wire has an impact on the viability of many proposed technological applications in sensing, storage, and logic operation

[123-130]. When a DW is driven by a magnetic field parallel to the long axis of the wire, the maximum wall speed is found to be function of, and capped by, the dimensions of the wire [131, 132]. In addition, the range of applied driving fields for which the wall motion is pure, fast translation is also limited by the wire geometry. Special effort has been performed in nanowires to enhance the DW speed. Thus, application of transverse magnetic field proposed in [126] allowed to improve slightly the DW speed till about 600 m/s at about 120 A/m.

One of the most interesting properties of glass-coated microwires is that related with the observation of spontaneous magnetic bistability in Fe-rich compositions [7-9]. This rectangular hysteresis loop was interpreted in terms of nucleation or depinning of the reversed domains inside the internal single domain and the consequent domain wall propagation [26].

Quite high DW velocities achieving up to 18 km/s have been reported in glass-coated microwires with few µm diameter [28]. Perfectly rectangular shape of the hysteresis loop has been related with a very high speed of such domain wall propagation. It is demonstrated by few methods that the remagnetization process of such magnetic microwire starts from the sample ends as a consequence of the depinning of the domain walls from closure domains and subsequent DW propagation from the closure domains [26, 133]. The magnetization process in axial direction runs through the propagation of the single head-to head DW. It is worth mentioning, that the micromagnetic origin of rapidly moving head-to head DW in microwires is still unclear, although there are evidences that this DW is relatively thick and has complex structure [131-133].

In the case of microwires the DW dynamics was measured using modified Sixtus Tonks [134] experiments, as described recently elsewhere [26, 28]. Usually the system consists of three coaxial pick-up coils (Fig.46). Additionally one end of the sample has been placed outside the solenoid in order to ensure domain wall nucleation always near one of the microwire ends. In this way in contrary to the classical Sixtus-Tonks experiments [132], we do not need the nucleation coils to nucleate the DW, since the closure domain wall already exists. The small closure domains are created at the ends of the wire in order to decrease the stray fields [7-9].

Then, DW velocity in this case can be estimated as:

$$v = \frac{l}{\Delta t} \qquad (14)$$

where $l$ is the distance between pick-up coils and $\Delta t$ is the time difference between the maximum in the induced *emf*.

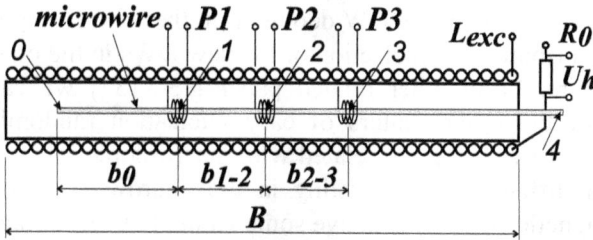

**Figure 47.** Schematic picture of the experimental set-up. Reprinted with permission from [136], M. Ipatov, V. Zhukova, A. K. Zvezdin and A. Zhukov, Mechanisms of the ultrafast magnetization switching in bistable amorphous microwires, *J. Appl. Phys.*, 106, 103902, 2009. Copyright [2009], AIP Publishing LLC. (Fig. 1).

In order to obtain the dependence of the DW velocity on magnetic field $v(H)$, it is necessary i) create a reverse domain in the certain, well-defined region of the sample, and ii) apply a stable magnetic field, $H$, of the required value along the wire axis. A set of coils (Fig. 47) was especially designed to fulfill these requirements. It consists of a long exciting coil $L_{exc}$ (with length $B$ of 140 mm, 10 mm in diameter) and tree pick-up coils $p_1$, $p_2$ and $p_3$ (2 mm long and 1 mm inner diameter) with distances $b_{1-2}$ and $b_{2-3}$ between coils of 27 mm. Each pick-up coil is connected to corresponding input of digital oscilloscope. In order to avoid the situation when the DW can start propagating while $H$ is still growing a single layered wounding of magnetizing solenoid with reduced number of turns has been used. The time of transient process is mainly defined by the inductance of the exciting coil (it also depends on slew rate of the signal source), which is proportional to the square of the number of turns, $N$. Therefore, reducing $N$ we shall reduce the transient time and increase the sweep rate, $dH/dt$. In order to achieve high enough magnetic field ($H{\sim}iN$) we used a power amplifier. The distance between the wire end and the first pick-up coil $b_0$ (approx 40 mm) was set in the way that the transient process has finished when the DW reaches the pick-up coil. In this way we achieved steady magnetic field, $H$, when the DW reaches the first coil $p_1$. The voltage drop $U_h$ on the resistor $R_0$ is proportional to the magnetic field in the exciting coil and is captured by channel 4 of the oscilloscope. In this way we control stability of magnetic field when measuring DW propagation. The described above technique guarantees that the DW velocity measurements are done at stable magnetic field.

In order to study the effect of magnetic field on single DW propagation, we need to control that this DW depins from the wire end (point 0) and to avoid contribution of nucleation of the new DWs in the other parts of the microwire. From other related paper [135-137] we can assume existence of the other centers of easy nucleation randomly located along the microwire that are related with macroscopic inhomogeneities (defects of different kind) existing in the microwire. In the case if applied magnetic field, $H$, is above some value, few DWs (one from the wire end and others from the reversed domains nucleated in the central part far from the wire's ends) can propagate simultaneously. In this case two pick-up coils set-up, previously employed elsewhere [26,28] probably does not allow revealing such situation.

As described above the preparation of glass-coated microwires involves simultaneous solidification of composite microwire consisting of ferromagnetic metallic nucleus inside the glass coating introduces considerable residual stresses inside the ferromagnetic metallic nucleus [7-9]. The strength of internal stresses is determined by thickness of glass coating and metallic nucleus diameter. This additional magnetoelastic anisotropy affects soft magnetic properties of glass-coated microwires. Consequently considerable attention has been paid to the effect of magnetoelastic anisotropy on DW dynamics in amorphous magnetically bistable microwires [137-140].

It is worth mentioning, that the magnetostriction constant, $\lambda_s$, in system $(Co_xFe_{1-x})_{75}Si_{15}B_{10}$ changes with x from $-5 \times 10^{-6}$ at x= 1, to $\lambda_s \approx 35 \times 10^{-6}$ at x≈0.2 [17-19]. Therefore, producing microwires with various Fe-Co rich compositions we were able to change the magnetostriction constant from $\lambda_s \approx 35 \times 10^{-6}$ for Fe-rich compositions $(Fe_{72.75}Co_{2.25}B_{15}Si_{10}$ and $Fe_{70}B_{15}Si_{10}C_5)$ till $\lambda_s \approx 10^{-7}$ for $Co_{56}Fe_8Ni_{10}Si_{10}B_{16}$ microwire. Additionally within each composition of metallic nucleus we also produced microwires with different ratio of metallic nucleus diameter and total diameter, $D$, i.e. with different ratios $\rho = d/D$. This allowed us to control residual stresses, since the strength of internal stresses is determined by ratio $\rho$ [7-9].

It is worth mentioning, that the magnetoelastic energy, $K_{me}$, is given by (13) depends on both total stress, $\sigma$, (i.e. on applied, $\sigma_a$, and internal, $\sigma_i$, stresses) and magnetostriction constant $\lambda_\sigma$.

In this way the effect of magnetoelastic contribution on DW dynamics has been experimentally studied controlling the magnetostriction constant, applied and/or residual stresses.

Hysteresis loops of a few studied microwires ($Fe_{70}B_{15}Si_{10}C_5$ and $Fe_{72.75}Co_{2.25}B_{15}Si_{10}$) with different metallic nucleus diameters and similar Fe-rich composition are shown in Fig. 48. As can be appreciated, considerable increasing of switching filed (from about 80 A/m till 700 A/m) is observed when ferromagnetic metallic nucleus diameter decreases from 15 till 1,4 μm (i.e. one order). At the same time, rectangular hysteresis loop shape is maintained even for smallest microwires diameters. Previously similar increasing of coercivity with decreasing the metallic nucleus diameters have been attributed to enhanced magnetoelastic energy arising from enhanced internal stresses when $\rho-$ ratio is small [7-9, 18]. Consequently, one of relevant parameters affecting strength of internal stresses and the magnetoelastic energy is $\rho$-ratio.

Usually it is assumed that domain wall (DW) propagates along the wire with a velocity:

$$v=S(H-H_0), \tag{15}$$

where $S$ is the DW mobility, $H$ is the axial magnetic field and $H_{0\ is}$ the critical propagation field.

Dependences of domain wall velocity, $v$, on magnetic field, $H$ for $Fe_{16}Co_{60}Si_{13}B_{11}$ and $Co_{41.7}Fe_{36.4}Si_{10.1}B_{11.8}$ amorphous microwires with the same $\rho$–ratio are shown in Fig. 49. In this case, the effect of only magnetostriction constant is that higher magnetostriction constant (in according to ref. [17-19] for $Co_{41.7}Fe_{36.4}Si_{10.1}B_{11}$ microwire $\lambda_\sigma \approx 25 \times 10^{-6}$ should be considered, while for $Fe_{16}Co_{60}Si_{13}B_{11}$ composition $\lambda_\sigma \approx 15 \times 10^{-6}$) results in smaller DW velocity at the same magnetic field and smaller DW mobility, $S$.

In order to evaluate the effect of $\rho$-ratio, i.e. effect of residual stresses on DW dynamics, we performed measurements of $v(H)$ dependences in the microwires with the same composition, but with different $\rho$–ratios.

Dependences of DW velocity on applied field for $Fe_{55}Co_{23}B_{11.8}Si_{10.1}$ microwires with different ratios are shown on Fig. 50. Like in Fig. 49, at the same values of applied field, $H$, the domain wall velocity is higher for microwires with higher $\rho$–ratio, i.e. when the internal stresses are lower [7-9].

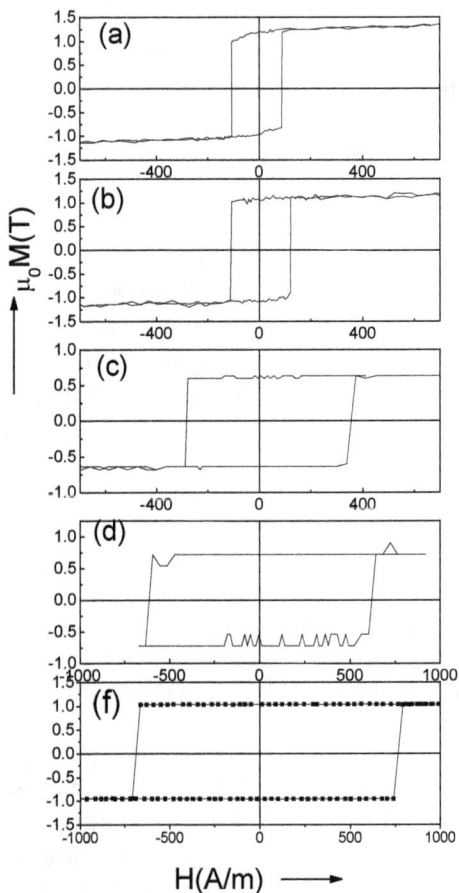

$H(A/m) \longrightarrow$

**Figure 48.** Hysteresis loops of Fe-rich amorphous microwires with the same sample length and different metallic nucleus diameter $d$ and total diameters $D$: $Fe_{70}B_{15}Si_{10}C_5$ microwires with $\rho = 0.63$; $d=15$ μm (a); $\rho= 0,48$; $d= 10,8$ μm (b); $\rho =0,26$; $d= 6$ μm (c); $\rho =0,16$; $d= 3$ μm (d) and of $Fe_{72.75}Co_{2.25}B_{15}Si_{10}$ microwire with $\rho= 0,14$; $d\approx 1,4$ μm (f). Reprinted with permission from [137] A. Zhukov, J. M. Blanco, M. Ipatov, A. Chizhik and V. Zhukova, Manipulation of domain wall dynamics in amorphous microwires through the magnetoelastic anisotropy, *Nanoscale Research Letters*, 7, (2012), 223, doi:10.1186/1556-276X-7-223. Copyright (2012) SpringerOpen (Fig. 1).

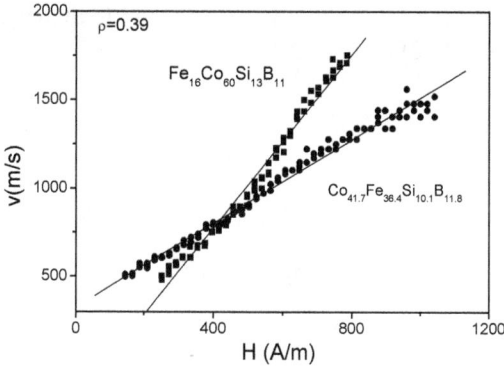

**Figure 49.** v (H) dependences for $Fe_{16}Co_{60}Si_{13}B_{11}$ and $Co_{41.7}Fe_{36.4}Si_{10.1}B_{11.8}$ microwires with $\rho$=0.39 . Reprinted with permission from [137] A. Zhukov, J. M. Blanco, M. Ipatov, A. Chizhik and V. Zhukova, Manipulation of domain wall dynamics in amorphous microwires through the magnetoelastic anisotropy, *Nanoscale Research Letters*, 7, (2012), 223, doi:10.1186/1556-276X-7-223. Copyright (2012) SpringerOpen (Fig. 2).

The other way to manipulate the magnetoelastic energy is to apply stresses during measurements. Fig. 50 shows *v(H)* dependences for $Co_{41.7}Fe_{36.4}Si_{10.1}B_{11.8}$ microwire ($\rho \approx 0.55$) measured under applied tensile stresses. Considerable decreasing of domain wall velocity, *v*, at the same magnetic field value, *H*, have been observed under application of applies stress. Additionally, increasing of applied stress, $\sigma_a$, results in decreasing of DW velocity.

Consequently in low magnetostrictive $Co_{56}Fe_8Ni_{10}Si_{10}B_{16}$ microwire DW velocity values achieved at the same values of applied field (see Fig. 51), are considerable higher (almost twice), than observed for microwires with higher magnetostriction constant (compare with Figs. 49-51).

As regarding experimentally observed *v(H)* dependences shown in Figs. 49-52, there are few typical features: linear extrapolation to zero domain wall velocity gives negative values of the critical propagation field, $H_o$. Such a negative value, previously reported for instance in Refs. [26, 140] has been explained in terms of the negative nucleation field of the reversed domain. In the case of amorphous microwires, the reversed domain already exists and does not need to be nucleated by the reversed applied magnetic field. Another typical feature is non-linearity of *v(H)* dependences at low-field region. Such deviations from linear dependence have been previously attributed to the domain wall interaction with the distributed defects and thermal fluctuations contribution [141, 142].

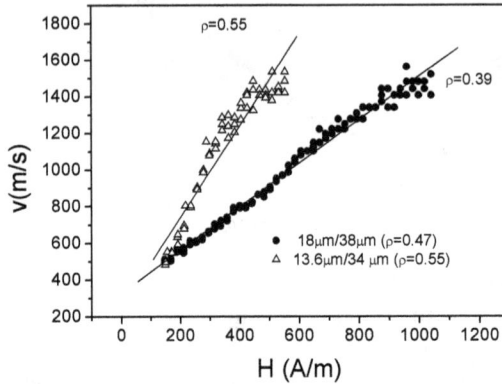

**Figure 50.** v(H) dependences for $Co_{41.7}Fe_{36.4}Si_{10.1}B_{11.8}$ microwires with different ratios $\rho$. Reprinted with permission from [139] A. Zhukov, J. M. Blanco, M. Ipatov, V. Rodionova, and V. Zhukova, Magnetoelastic Effects and Distribution of Defects in Micrometric Amorphous Wires, *IEEE Transactions on Magnetics*, Vol. 48, Issue: 4, (2012) 1324-1326. Copyright (2012) IEEE (Fig. 4).

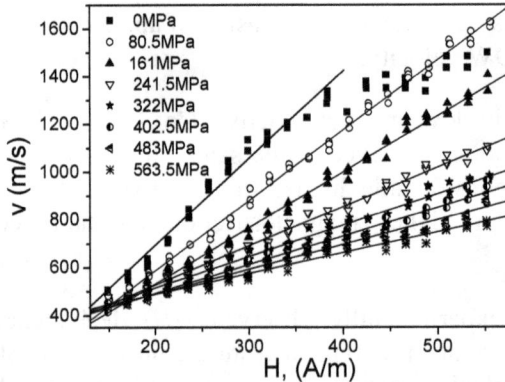

**Figure 51.** *v(H)* dependences for $Co_{41.7}Fe_{36.4}Si_{10.1}B_{11.8}$ microwires (d≈ 13.6μm, D≈ 24,6μm, ρ≈ 0.55) measured under application of applied stresses, $\sigma_a$. Reprinted with permission from [137]. A. Zhukov, J. M. Blanco, M. Ipatov, A. Chizhik and V. Zhukova, Manipulation of domain wall dynamics in amorphous microwires through the magnetoelastic anisotropy, *Nanoscale Research Letters*, 7, (2012) 223, doi:10.1186/1556-276X-7-223 Copyright (2012) SpringerOpen (Fig. 4).

The domain wall dynamics in viscous regime is determined by a mobility relation (15), where $S$ is the domain wall mobility given by:

$$S=2\mu_0 M_\sigma/\beta, \qquad (16)$$

where $\beta$ is the viscous damping coefficient, $\mu_0$ is the magnetic permeability of vacuum. Damping is the most relevant parameter determining the domain wall dynamics. Various contributions to viscous damping $\beta$ have been considered and two of them are generally accepted [137-139]:

- micro-eddy currents circulating nearby moving domain wall are the more obvious cause of damping in metals. However, the eddy current parameter $\beta_\varepsilon$ is considered to be negligible in high-resistive materials, like thin amorphous microwires.

The second generally accepted contribution of energy dissipation is magnetic relaxation damping, $\beta_\rho$, related to a delayed rotation of electron spins. This damping is related to the Gilbert damping parameter and is inversely proportional to the domain wall width $\delta_\omega$ [141,142],

$$\beta_r \approx \alpha M_s /\gamma\Delta \approx M_s(K_{me}/A)^{1/2}, \qquad (18)$$

where $\gamma$ is the gyromagnetic ratio, $A$ is the exchange stiffness constant, $K_{me}$ is the magnetoelastic anisotropy energy given by (13).

Consequently, the magnetoelastic energy can affect domain wall mobility, $S$, as we experimentally observed in few Co-Fe-rich microwires.

Considering the aforementioned, we can suggest, that DW velocity, $v$, should decrease with stress and magnetostriction constant increasing and if only the magnetoelastic energy affects the DW dynamics, $v$ should show an inverse square root dependence on stress or magnetostriction. Therefore, we tried to evaluate the $v(\sigma_{app})$ dependence. An example of $v(\sigma_{app})$ dependence, obtained for $Fe_{55}Co_{23}B_{11.8}Si_{10.2}$ microwires (d=13.2 µm; D=29.6 µm), is shown in Fig. 52 a. Qualitatively, we observed decreasing of DW velocity, $v$, with applied stresses, $\sigma_{app}$. In order to evaluate, if obtained dependence $v(\sigma_{app})$ fits inverse square root dependence on applied stress, $\sigma_{app}$, we expressed obtained dependences as $\sigma_{app}(v^{-2})$. From Fig. 52 b we can conclude, that obtained $v(\sigma_{app})$ dependences cannot be described by single $v(\sigma_{app}^{-1/2})$ dependence. On the other hand, at high enough $\sigma_{app}$ observed $v(\sigma_{app})$ dependence probably can be fitted by two $v(\sigma_{app}^{-1/2})$.

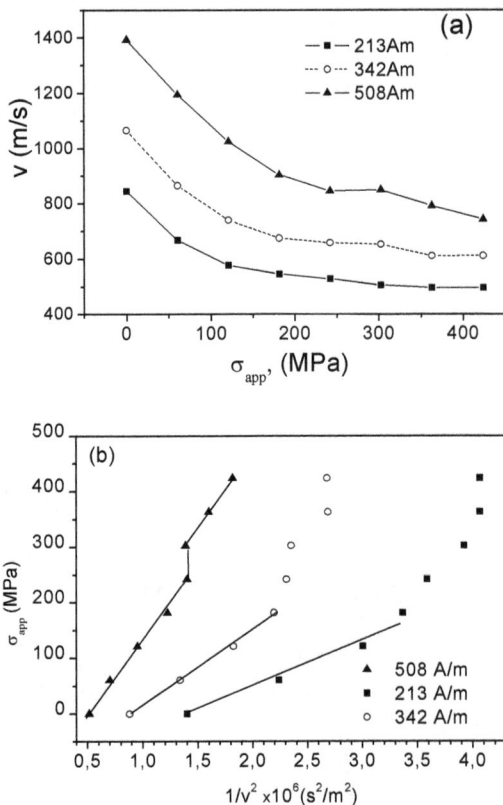

**Figure 52.** $v(\sigma_{app})$ dependences of $Fe_{55}Co_{23}B_{11.8}Si_{10.2}$ microwires (d=13,2 μm; D=29,6 μm) (a) and $\sigma_{app}$ $(1/v^2)$ dependence (b). Reprinted with permission from [137]. A. Zhukov, J. M. Blanco, M. Ipatov, A. Chizhik and V. Zhukova, Manipulation of domain wall dynamics in amorphous microwires through the magnetoelastic anisotropy, *Nanoscale Research Letters*, 7, (2012), 223, doi:10.1186/1556-276X-7-223. Copyright (2012) SpringerOpen (Fig. 6)

It is worth mentioning, that systematic analysis of mechanisms of DW dynamics in thicker (with diameters between 30 and 120 μm) magnetostrictive amorphous wires without glass has been performed in Ref. 143 on the basis of bubble domain dynamics. The systematic analysis method in this paper is also a strong basis for considering domain propagation dynamics in glass-covered thinner magnetostrictive amorphous wires. Main assumptions on domain wall configuration in thicker wires have been performed considering, that the DW length, $l$, is much more than its radius, $r$ $(r/l << 10^{-3})$.

Recently the attempt to extend the analysis performed in ref. [143] to thinner glass-coated microwires (typically with diameter of the order of 10 μm) with strong internal stresses induced by the glass-coating [144]. Particularly analyzing the voltage peaks forms and experimental data on DW dynamics we demonstrated that a very high DW mobility observed in magnetically bistable amorphous microwires with a diameter of about 10 μm can be associated with elongated domain shape. The experimental results can be explained in terms of the normal mobility with respect to the domain surface, which is reduced by a factor representing the domain aspect ratio estimated to be in the range of 300 for considered wire samples. On the other hand, experimental data on DW dynamics in thin microwires and analysis of the voltages on pick-up coils show, that generally the structure of propagating DW is far from abrupt and quite complex [131, 132]. Thus, the characteristic width of the head to head DW, $\delta$, depends on many factors, such as applied magnetic field, $H$: at $H = 60$A/m $\delta \approx 65d$, while at $H = 300$ A/m, $\delta \approx 40d$. Additionally, depends on magnetic anisotropy constant, $K$, being $\delta/d \approx 13.5$ for $K = 10^4$ erg/cm$^3$, $\delta/d \approx 20$ for $K = 5 \times 10^3$ erg/cm$^3$, $\delta/d = 30$–$34$ for $K = 2 \times 10^3$ erg/cm$^3$ and $\delta/d = 40$–$50$ for $K = 10^3$ erg/cm3, respectively [131].

Regarding aforementioned, it is interesting to compare the velocity of DW propagation in thinnest microwire with the values observed in submicrometric planar nanowires reported elsewhere [145]. The DW velocity in thin microwire is ranging between 700 and 850 m/s (Fig. 7a), which is still higher than for the same range of magnetic field as –compared with submicrometric nanowires (maximum v $\approx$110 m/s at 700 A/m) reported elsewhere [145].

On the other hand for such elevated magnetic fields (1000-1500 A/m) the domain wall velocity, $v$, is significantly lower than for thicker wires of the same composition with lower ρ-ratio. For comparison $v(H)$ dependence for $Fe_{74}Si_{11}B_{13}C_2$ microwire with similar composition with metallic nucleus $d$ and total $D$ diameters 12.0/15.8 is presented in Fig. 53. As can be deduced from comparison of DW dynamics thicker $Fe_{74}Si_{11}B_{13}C_2$ microwire at maximum achieved magnetic field (about 280 A/m) presented double higher velocity as-compared with

$Fe_{72.75}Co_{2.25}B_{15}Si_{10}$ amorphous microwire with metallic nucleus diameter, $d$, of 2.8 μm and total diameter $D \approx 9$ μm (Fig. 53).

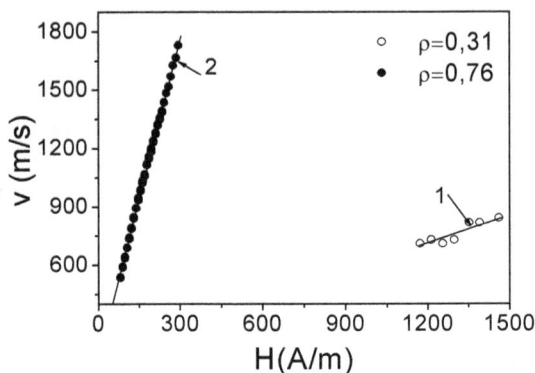

**Figure 53.** $v(H)$ dependence for $Fe_{72.75}Co_{2.25}B_{15}Si_{10}$ amorphous microwires with metallic nucleus diameter, $d$, of 2,8 μm and total diameter $D \approx 9$μm (1) and $d=12.0/D=15.8$ (2). Reprinted with permission from [137]. A. Zhukov, J. M. Blanco, M. Ipatov, A. Chizhik and V. Zhukova, Manipulation of domain wall dynamics in amorphous microwires through the magnetoelastic anisotropy, *Nanoscale Research Letters*, 7, (2012), 223, doi:10.1186/1556-276X-7-223. Copyright (2012) SpringerOpen (Fig. 7 b).

As regarding observed differences on $v(H)$ dependences one should consider enhanced magnetoelastic energy for $Fe_{72.75}Co_{2.25}B_{15}Si_{10}$ amorphous microwire, since ratio $\rho=d/D$, determining strength of internal stresses[14,15] for thin $Fe_{72.75}Co_{2.25}B_{15}Si_{10}$ is $\rho \approx 0.31$, while for thicker $Fe_{74}Si_{11}B_{13}C_2$ microwire $\rho \approx 0.56$.

This also reflected by the change of the shape of the voltage induced in the pick-up coil surrounding microwires under tensile stress application (see Fig. 54 for the $Fe_{74}B_{13}Si_{11}C_2$ microwire with $\lambda_\sigma \approx 35x10^{-6}$). As-prepared microwires exhibit quite sharp voltage peaks induced in the pick-up coil associated with fast magnetization switching with the half-width of the peak about 3 μsec. Applying tensile stress, the gradually increases and at 260 MPa achieves about 8 μsec. Such increasing of the half-width reflects decreasing of DW velocity under tensile stress application.

In summary, we experimentally observed that manipulating the magnetoelastic energy through application of tensile stress, changing the magnetostriction constant and internal stresses of studied microwires we significantly affected the domain wall dynamics in magnetically bistable microwires. Considering aforementioned we assume that in order to achieve higher DW propagation velocity at the same magnetic field and enhanced DW mobility special attention should be paid to decreasing of magnetoelastic energy.

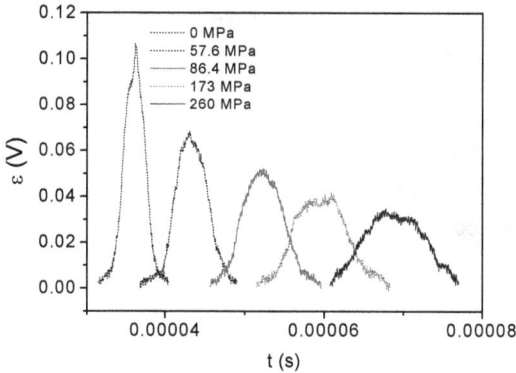

**Figure 54.** Change of shape of the voltage from pick-up coil under tensile stress application in $Fe_{74}B_{13}Si_{11}C_2$ (d≈14.6 μm, D≈21.8 μm, ρ≈0.67) microwires. Reprinted with permission from [137]. A. Zhukov, J. M. Blanco, M. Ipatov, A. Chizhik and V. Zhukova, Manipulation of domain wall dynamics in amorphous microwires through the magnetoelastic anisotropy, *Nanoscale Research Letters*, 7, (2012), 223, doi:10.1186/1556-276X-7-223. Copyright (2012) SpringerOpen (Fig. 8).

## 5.3. Role of Defects

Recently have been observed that microwires, like any materials present different kind of defects that can be detected through the measurements of the local nucleation fields of reversed domains [134, 136].

Considering a number of unusual effects found studying DW propagation in amorphous microwires we tried to reveal contribution of local defects on peculiarities of domain wall propagation and correlation of linearity of *v(H)* dependence with defects existing in amorphous microwires. A comparative study of single domain wall dynamics and local nucleation fields in Fe-rich amorphous glass-coated microwires has been performed.

To detect the possible nucleation and subsequent propagation of several DWs, the three pick-up coils setup described above has been employed.

For measurements of local nucleation fields the special set-up has been developed. To create a small domain with opposite magnetization far from the wire ends a short micro coil that produces a magnetic field strong enough to nucleate a reversed domain has been used. Thus, a pair of head to head and tail to tail DWs can be created in a controllable manner avoiding the nucleation of the DWs at the wire end, as it usually occurs in the experiments [135]. The magnetic field value necessary for nucleation of the reversed domain gives us a local value of the wire nucleation field in various points along the wire length. This measurement can be made under applied tension stress created by a small load attached to the bottom wire end.

In this way the remagnetization dynamics in magnetically bistable Fe-rich microwire has been studied.

The distributions of the $H_n(x)$ for Fe-rich samples are shown in Fig. 55. A number of dip holes on $H_n(x)$ curves for both samples, attributed previously to the positions of localized defects existing within the microwire [136, 146]. The overall minimum, $H_N$, observed for both microwires from $H_n(x)$.has been compared with the $v(H)$ dependence of the same samples.

Fig. 56 shows the measured dependences of DW velocity on applied magnetic field in the same bistable Fe-rich amorphous glass-coated microwires.

The linear $v(H)$ regime has been observed in both samples till 294 and 340 A/m respectively for samples 1 and 2. The DW mobility $S$ about 5 m$^2$/As and maximum DW velocity $v$ of 1.7 km/s have been observed. The mechanism of such ultrafast magnetization switching in second regime of the sample magnetization reversal is considered below in details.

It was assumed that such drastic change of the remagnetization process and deviation from linear $v(H)$ dependence is caused by possible nucleation and consequent growing of additional reversed domain with lowest local nucleation field, $H_N$. This new reverse domain can be located at any place inside the sample.

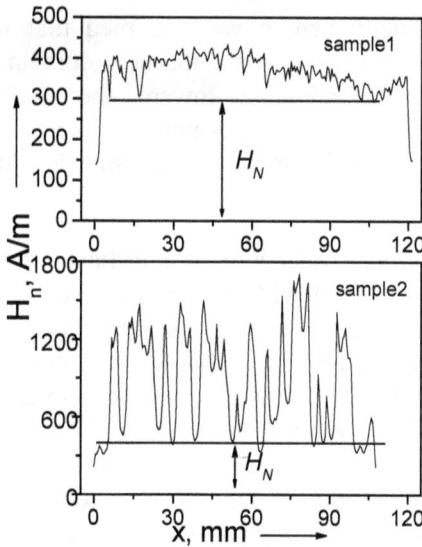

**Figure 55.** Distribution of local nucleation fields measured in $Fe_{74}Si_{11}B_{13}C_2$ (sample 1) and $Fe_{75}Si_{12}B_9C_4$ (sample 2) microwires. Reprinted with permission from [136]. M. Ipatov, V. Zhukova, A. K. Zvezdin and A. Zhukov, Mechanisms of the ultrafast magnetization switching in bistable amorphous microwires, *J. Appl. Phys.*, 106, 103902, 2009. Copyright [2009], AIP Publishing LLC. (Fig. 3 b).

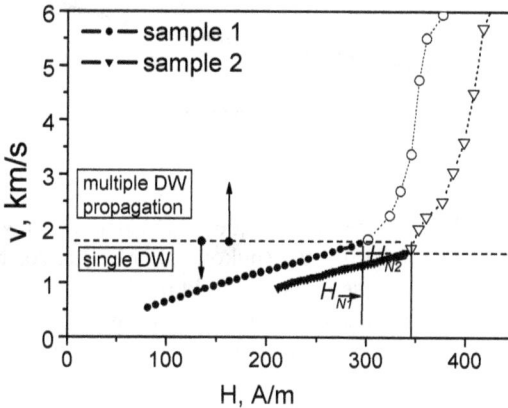

**Figure 56.** Magnetic field dependences of domain wall velocity measured in $Fe_{74}Si_{11}B_{13}C_2$ (sample 1) and $Fe_{75}Si_{12}B_9C_4$ (sample 2) microwires. Reprinted with permission from [136]. M. Ipatov, V. Zhukova, A. K. Zvezdin and A. Zhukov, Mechanisms of the ultrafast magnetization switching in bistable amorphous microwires, *J. Appl. Phys.*, 106, 103902, 2009. Copyright [2009], AIP Publishing LLC. (Fig. 3 a).

Considering aforementioned, it was assumed that when the applied magnetic field has reached the $H_N$, the new domain is nucleated and two more DW starts to propagate towards the wire's ends. As it was noted above, in such situation it is not possible to measure correctly single DW velocity and we can only consider the effective DW velocity.

It should be mentioned, that similar non-linear dependence of DW velocity on magnetic field with abrupt increasing of DW velocity at elevated magnetic fields can be observed in most of amorphous microwires (see Fig. 57 for $Co_{56}Fe_8Ni_{10}Si_{11}B_{16}$ microwire). In particular case of $Co_{56}Fe_8Ni_{10}Si_{11}B_{16}$ microwire, the magnetostriction constant, $\lambda_\sigma$, is small and positive (about $10^{-6}$). Therefore, lower coercivity (about 10 A/m) and DW propagation at lower field are observed in $Co_{56}Fe_8Ni_{10}Si_{11}B_{16}$ microwire. High DW velocity, $v$, ($v \approx$ 3000 m/s) before abrupt jump on $v$ $(H)$ dependence is observed.

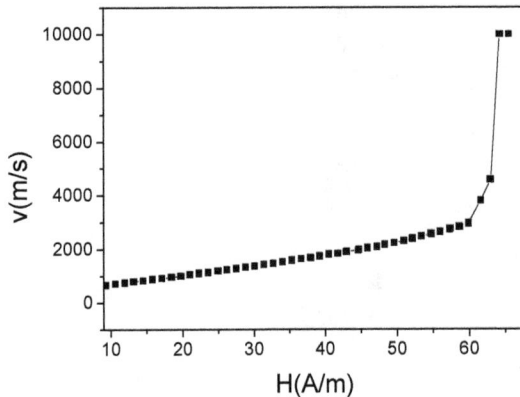

**Figure 57.** v (H) dependence for $Co_{56}Fe_8Ni_{10}Si_{11}B_{16}$ microwire. Reprinted with permission from ref. [150] V. Zhukova, J. M. Blanco, M. Ipatov and A. Zhukov, Effect of transverse magnetic field on domain wall propagation in magnetically bistable glass-coated amorphous microwires, *J. Appl. Phys.*, 106, 113914, 2009. Copyright [2009], AIP Publishing LLC. (Fig. 6 b).

Consequently, it is worth distinguishing an interesting mechanism of magnetization reversal in magnetically bistable microwires, related with the microwires defects. Existence of defects can lead to considerable acceleration of the microwire remagnetization process. On the other hand, neglecting of this factor can result in exaggerated estimation of DW velocity from Sixtus-Tonks experiment when using just 2 pick-up coils. Consequently at magnetic field above $H_N$, the contribution of defects can be essential. Appearance of these additional domains at $H > H_N$ can accelerate remagnetization switching resulting in higher effective DW velocity.

Obviously, this process, which can be nominated as tandem remagnetization, results in significant decrease of the magnetization switching time and acceleration of magnetization switching in magnetically bistable microwires. Proposed mechanism of ultrafast magnetization switching can explain non-linearity of $v(H)$ dependences and ultra-fast DW propagation reported in magnetically bistable microwires [136, 146].

The essence of this process is clearly shown in Fig. 58. There left front of propagating head-to-head domain wall *dw2* moves toward the *dw1*. Finally these two reversed domains clamping and the right front, *dw3*, becomes the unique. Obviously, this process, which can be nominated as tandem remagnetization, results in significant decrease of the magnetization switching time and acceleration of magnetization switching in magnetically bistable microwires. Proposed mechanism of ultrafast magnetization switching can explain non-linearity of $v(H)$ dependences and ultra-fast DW propagation reported in magnetically bistable microwires [136].

Local nucleation field is typical for a given sample. Fig. 59 shows the distribution of the local nucleation fields $H_N$ along the sample length, $x$, measured using short movable magnetizing coil placed far from the samples ends and moving in the opposite directions. In order to distinguish we have marked one of the microwire's ends as "end1" and other as "end2". Black solid line shows the distribution of the local nucleation fields measured using moving short magnetization coil between the end1 to the end2. Red one (dotted line) is for the movement in the opposite direction. As we can observe from Fig. 59, the nucleation field near wires ends is considerably smaller. This characteristic feature is responsible for the origin of large Barkhausen jump, when fast switching of the magnetization runs by depinning and consequent fast DW propagation from one of the wire ends. One can see also from $H_n(x)$ distributions, that the dip holes keep the same position moving the short magnetizing coil from end 1 to end 2 and back, i.e. this nucleation field profile is specific for a given microwire sample.

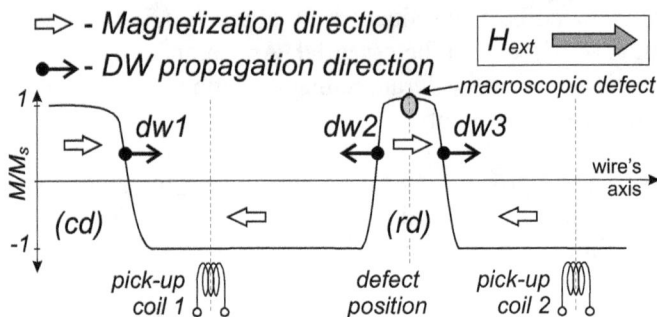

**Figure 58.** Magnetization switching through tandem mechanism, *cd* – closure domain, *rd*-reversed domain appeared within the microwire. Reprinted with permission from [136]. M. Ipatov, V. Zhukova, A. K. Zvezdin and A. Zhukov, Mechanisms of the ultrafast magnetization switching in bistable amorphous microwires, *J. Appl. Phys.*, 106, 103902, 2009. Copyright [2009], AIP Publishing LLC. (Fig. 4).

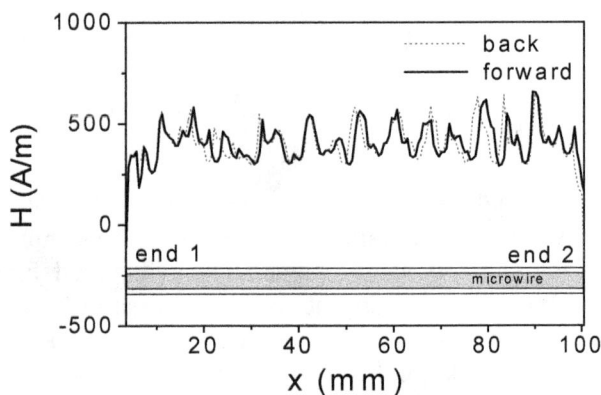

**Figure 59.** Distribution of the local nucleation fields measured in different directions along the sample of magnetically bistable amorphous $Fe_{74}B_{13}Si_{11}C_2$ microwire (see the arrows identification in the text). Reprinted with permission from [140]. V. Zhukova, J.M. Blanco, V. Rodionova, M. Ipatov, A. Zhukov, Fast magnetization switching in Fe-rich amorphous microwires: Effect of magnetoelastic anisotropy and role of defects, *J. Alloys Comp.*, 586 (SUPPL. 1), (2014), pp. S287-S290, Copyright (2014), with permission from Elsevier. (Fig. 3).

In Fig. 60 a correlation between the $H_n(x)$ and DW dynamics in $Fe_{74}B_{13}Si_{11}C_2$. microwire is demonstrated. Fig. 60a shows $H_n(x)$ dependence. The deepest minimum is observed between 1 and 2 pick-up coils at $H_n \approx 170$ A/m. Fig. 60 b represents dependence of DW velocity, $v$, on applied magnetic field, $H$, measured by pair of coils 1-2 ($v_{1-2}$) and 2-3 ($v_{2-3}$). These dependences exhibit significant difference: both $v_{1-2}(H)$ and $v_{2-3}(H)$ at low fields show linear growth with $H$, but at $H \approx 190$ A/m we observed abrupt jump on $v_{1-2}(H)$, while $v_{2-3}$ continue growing with $H$. The field of jump on $v_{1-2}(H)$ corresponds to the minimum nucleation field $H_n \approx 170$ A/m observed in Fig. 60 b.

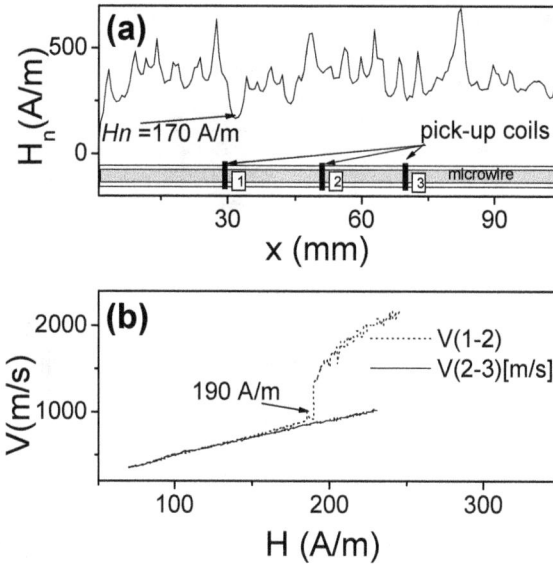

**Figure 60.** Correlation of local nucleation fields distribution (a) and $v(H)$ dependences in magnetically bistable amorphous $Fe_{74}B_{13}Si_{11}C_2$ microwire ($d \approx 19.4$, $D \approx 26.6$, $\rho \approx 0.73$). 1, 2, 3 are the positions of the pick-up coils. Reprinted with permission [140]. V. Zhukova, J.M. Blanco, V. Rodionova, M. Ipatov, A. Zhukov, Fast magnetization switching in Fe-rich amorphous microwires: Effect of magnetoelastic anisotropy and role of defects, *J. Alloys Comp.*, 586 (SUPPL. 1), (2014), pp. S287-S290, Copyright (2014), with permission from Elsevier. (Fig. 4).

Consequently, the microwires' inhomogeneities sufficiently affect the remagnetization process [136, 146], as was observed through the measurements of the distribution of the local nucleation fields, $H_N$, along the sample length, L. The origin of the defects might be related to stress inhomogeneities, shape irregularities, oxides etc. It is assumed, that at least some of these defects might have a magnetoelastic origin and, therefore, might be affected by heat treatment.

## 5.4. DW Dynamics Manipulation

It is assumed, that at least some of these defects might have a magnetoelastic origin and, therefore, might be affected by heat treatment. Recently some interesting results on effect of annealing on DW dynamics, coercivity and local nucleation fields distribution have been reported [147].

Dependences of the coercivity, $H_C$, on annealing time, $t_{AN}$, for the $Fe_{74}B_{13}Si_{11}C_2$ microwire measured at two different temperatures, $T_{AN}$ are shown in Fig. 61.

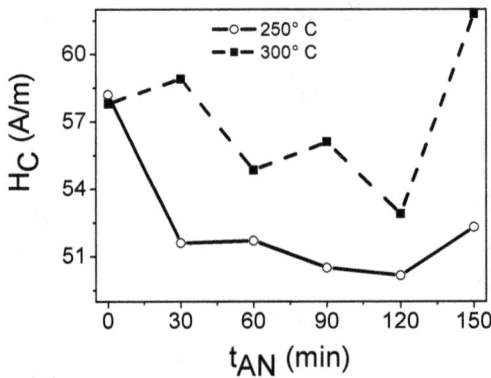

**Figure 61.** Dependences of the coercive force of the samples, annealed at different temperatures, on annealing time for $Fe_{74}B_{13}Si_{11}C_2$ microwire. Reprinted with permission from [147]. K. Chichay, V. Zhukova, V. Rodionova, M. Ipatov, A. Talaat, J. M. Blanco, J. Gonzalez and A. Zhukov, Tailoring of domain wall dynamics in amorphous microwires by annealing, *J. Appl. Phys.*, 17A318, (2013). Copyright [2013], AIP Publishing LLC (Fig. 1).

The considerable difference in $H_C(t_{AN})$ for $T_{AN}$ = 250 and 300 °C observed in the $Fe_{74}B_{13}Si_{11}C_2$ microwire might be explained by considering the main contribution of stress relaxation at lower $T_{AN}$ and $t_{AN}$, and the increasing influence of first crystallization processes at higher $T_{AN}$ and elevated $t_{AN}$.

The dependences for DW velocity between coils 2-3, $V(2-3)$, for the $Fe_{66.7}Cr_{11.4}B_{12}Si_9Ni_{0.9}$ microwires, and between coils 1-3, $V(1-3)$, for the $Fe_{74}B_{13}Si_{11}C_2$ microwires are presented in the Fig. 62. We observed typical almost linear $v(H)$ dependences in both as-prepared samples.

After heat treatments at both temperatures we observed extending of the magnetic field range where linear $v(H)$ dependence takes place (which corresponds to single DW propagation regime) and increasing of the DW velocity. At longest annealing time (above 120 min) both parameters (DW velocity and range of magnetic field for linear $v(H)$ dependence) are almost insensible on $t_{AN}$, exhibiting tendency to saturation.

In order to illustrate better the change in DW dynamics induced by annealing, we plotted the DW velocity value measured at the same magnetic field after different heat treatments as a function of annealing time, choosing the magnetic field values at which linear $v(H)$ dependence has been observed after all heat treatments. For the $Fe_{66.7}Cr_{11.4}B_{12}Si_9Ni_{0.9}$ sample, we selected $H = 300$ A/m and for the $Fe_{74}B_{13}Si_{11}C_2$ microwire, $H = 130$ A/m. Dependences of DW velocity on annealing time obtained for both studied samples are presented in Fig. 62 (a, b). We can observe a clear tendency of increasing of DW velocity with increasing of $t_{AN}$ at fixed $H$ values. The increase of the velocity observed in the $Fe_{66.7}Cr_{11.4}B_{12}Si_9Ni_{0.9}$ sample annealed at $T_{AN} = 250°C$ (Fig. 62 a) is less pronounced in comparison with the $Fe_{74}B_{13}Si_{11}C_2$ microwire annealed at $T_{AN} = 300°C$ (Fig. 62 b). Moreover, for the as-prepared samples, the difference for the velocities $V_{1-2}$, $V_{2-3}$ and $V_{1-3}$ is larger than for annealed samples with different $t_{AN}$.

Local nucleation field distributions in as-prepared and annealed ($T_{AN} = 250$ °C, for $t_{AN} = 150$ minutes) $Fe_{66.7}Cr_{11.4}B_{12}Si_9Ni_{0.9}$ microwires, as well as in as-prepared and annealed at $T_{AN} = 300°C$ $Fe_{74}B_{13}Si_{11}C_2$ microwires, are shown in Fig. 63 (a, b). As can be appreciated from Fig. 63 (a, b), a decrease of oscillation amplitudes in $H_N(L)$ dependences takes place after annealing.

Observed decreases of the amplitude of local nucleation fields, increasing DW velocity, and the extending of the magnetic field range of single-DW propagation regime could be attributed to stress relaxation after annealing. Moreover, extending the magnetic field range for a single-DW propagation regime is in agreement with a considerable increase of the local nucleation fields (from 20 to 40 %), previously reported for $Fe_{74}B_{13}Si_{11}C_2$ amorphous glass-coated microwire after annealing [148].

It is worth mentioning that there is a correlation between the amplitude of the $H_N(L)$ oscillations on local nucleation field distributions, and the type of DW propagation [149].

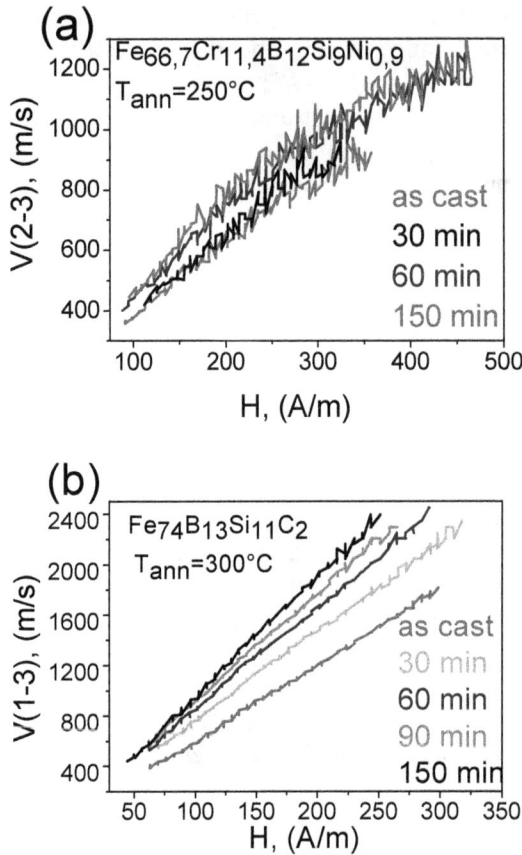

**Figure 62.** Dependences of DW velocity on magnetic field for as-prepared samples (a) for $Fe_{66.7}Cr_{11.4}B_{12}Si_9Ni_{0.9}$ microwire annealed at $T_{AN}$ = 250 °C for a 30, 60, 150 min and (b) for $Fe_{74}B_{13}Si_{11}C_2$ microwire annealed at $T_{AN}$ =300 °C for a 30, 60, 90, 150 min. Reprinted with permission from [147]. K. Chichay, V. Zhukova, V. Rodionova, M. Ipatov, A. Talaat, J. M. Blanco, J. Gonzalez and A. Zhukov, Tailoring of domain wall dynamics in amorphous microwires by annealing, *J. Appl. Phys.*, 17A318 (2013) Copyright [2013], AIP Publishing LLC (Fig. 2).

The accelerated DW propagation in microwires with higher amplitude $H_N(L)$ oscillations, and uniform DW propagation for smoother $H_N(L)$ distribution [149] have been observed. As observed in Fig. 63 decreasing differences between $V_{1-2}$, $V_{2-3}$ and $V_{1-3}$ values after annealing correlate well with recently reported correlations of the type of DW propagation with the defect density.

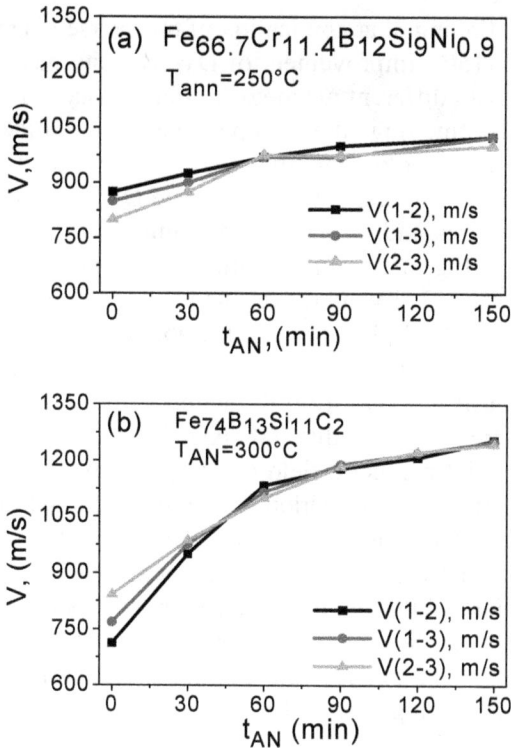

**Figure 63.** Dependences of DW velocity on annealing time (a) for $Fe_{66.7}Cr_{11.4}B_{12}Si_9Ni_{0.9}$ sample annealed at $T_{AN} = 250°C$ (plotted for fixed magnetic field value $H = 300$ A/m) and (b) for $Fe_{74}B_{13}Si_{11}C_2$ microwires annealed at $T_{AN} = 300°C$ (plotted for $H = 130$ A/m). Reprinted with permission from [147]. K. Chichay, V. Zhukova, V. Rodionova, M. Ipatov, A. Talaat, J. M. Blanco, J. Gonzalez and A. Zhukov, Tailoring of domain wall dynamics in amorphous microwires by annealing, *J. Appl. Phys.*, 17A318, (2013). Copyright [2013], AIP Publishing LLC (Fig. 3).

Consequently we can assume that at least a proportion of the samples' inhomogeneities, manifested through the oscillations of the local nucleation fields and limiting single DW regime, have an origin related to inhomogeneities in the internal stress distribution. Additionally, from the results presented above, we may assume two possible mechanisms of influence on the defects on DW dynamics: DW nucleation, and the pinning of propagating DWs. Annealing of microwires, giving rise to internal stress relaxation and/or release of some defect, can enhance the DW velocity at a given magnetic field, as well as allow us to extend the range of magnetic fields in which a single DW propagation takes place.

It is worth mentioning, that application of transverse magnetic field can affect the DW [150]. Improvement of DW velocity has been observed in microwires with different magnetostriction constant (Fig. 64). On the other hand, by moving the sample with the holder inside the magnetization coil and/or applying the bias field at different angles with respect to the wire axis, we can activate DW propagation from the opposite microwires end. Examples of manipulation of DW dynamics are presented in Fig. 65, where a temporal dependence of the voltage peaks in the coils 1, 2, 3 wound over the 10 cm long wire and separated by 27 mm is seen. It can be seen that the application of a bias magnetic field, $H_b$, at an angle of $\Theta \sim 5°$ with respect to the wire axis can activate the propagation of the DW from the wire end outside the solenoid. In this case $H_b$ creates an additional axial magnetic field component. Therefore the axial magnetic field at any point of the microwire must be considered as the superposition of the applied axial magnetic field, $H$, and the axial projection of $H_b$. At certain conditions (values of $H$, $H_b$ and the angle $\Theta$ between the axial direction and the applied bias field) the superposition of the applied axial magnetic field, $H$, and the $H_b$ axial projection is largest at the end of the sample outside the magnetizing. At this moment the DW propagation from the sample end outside the magnetizing coil will be activated (see a schematic picture in Fig. 66).

As can be appreciated from Figs. 65 (a-d), by changing value of $H_b$ (and consequently the axial $H_b$ projection) we can activate in a controllable way the DW propagation from the opposite end of the sample and induce a DW collision at various locations in the sample. If this collision takes place in the vicinity of a pick-up coil we will observe an increase of the voltage peak amplitude. Consequently we are able to engineer the DW collisions between two moving DWs in different places of the magnetic microwire by controlling an applied bias field. As observed in Fig. 60, defects existing in microwire limit the DW velocity, because new DW can nucleate defects. The macroscopic inhomogeneities (defects) existing in the microwire observed through the DW nucleation field fluctuations along the sample length is related to. This gives rise to a multi-domain remagnetization process when a few DWs will propagate simultaneously. Manipulating the bias field we able to release the domain walls at targeted locations in the microwire.

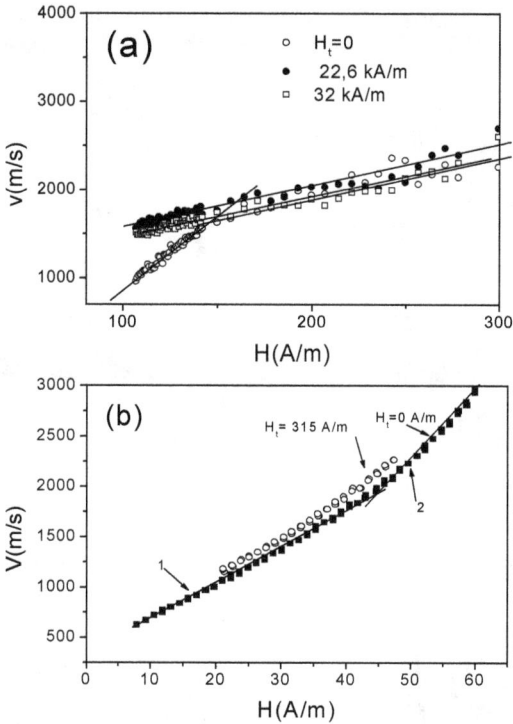

**Figure 64.** Magnetic field dependence of domain wall velocity in (a) $Fe_{69}Si_{10}B_{15}C_6$ and (b) $Co_{56}Fe_8Ni_{10}Si_{11}B_{16}$ measured without transverse magnetic field and under applied magnetic field. Reprinted with permission from [150]. V. Zhukova, J. M. Blanco, M. Ipatov and A. Zhukov, Effect of transverse magnetic field on domain wall propagation inmagnetically bistable glass-coated amorphous microwires, *J. Appl. Phys.*, 106, 113914, 2009. Copyright [2009], AIP Publishing LLC (Figs. 3, 4).

From analyzing the data in Fig. 65 (c) we find that at $H_b \approx 252$ A/m the DW collision takes place near the position of the 2nd coil. In this case the height of the signal from coil 2 drastically increases (Fig. 65 b, c). Additionally, when the $H_b$ value is high enough, the DW from the opposite end of the sample arrives at coil 1 simultaneously with the DW propagating from the nearest end of the sample (Fig. 65 d). In this case we observed the DW collision by coil 1 (Fig. 65 d). Similarly, when we moved the sample inside the magnetizing coil we were able to activate DW propagation from one side of the sample or from the other one depending on the position of the sample inside the solenoid. If at a certain position of the sample inside the magnetizing coil the effective axial magnetic field acting on the opposite end of the sample is higher than $H_n$, DW propagation can be observed from the opposite end of the sample.

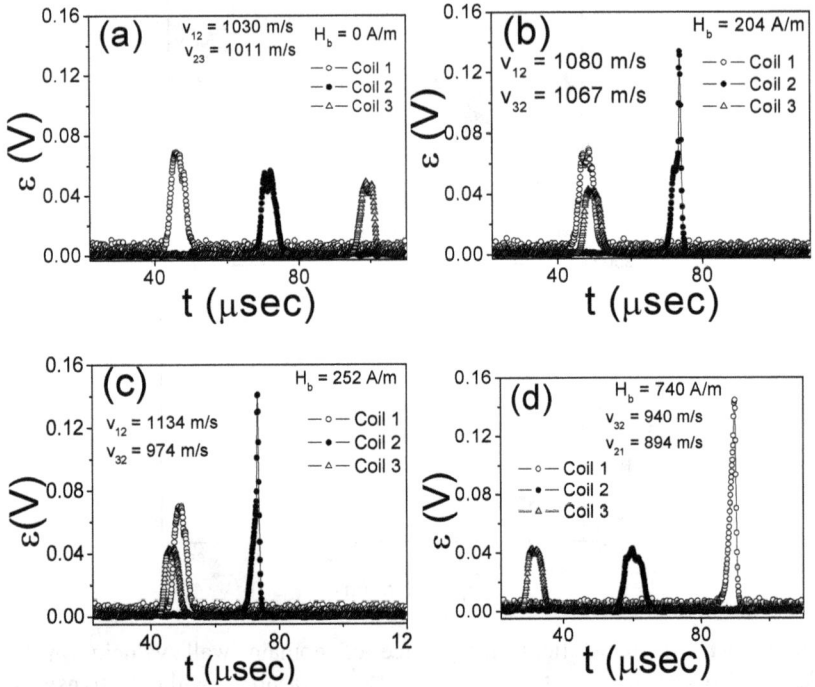

**Figure 65.** Voltage peaks induced by the propagating DW in the 3 pick-up coils in $Fe_{74}B_{13}Si_{11}C_2$ ($d\approx14.6$ μm, $D\approx21.8$ μm, $\rho\approx0.67$) microwires measured at H=140A/m with different values of bias magnetic field, $H_b$. Reprinted with permission from [142]. Zhukov, J. M. Blanco, A. Chizhik, M. Ipatov, V. Rodionova, and V. Zhukova, Manipulation of domain wall dynamics in amorphous microwires through domain wall collision, *J. Appl. Phys.*, 114, 043910, (2013). Copyright [2013], AIP Publishing LLC (Fig. 4).

Summarizing, the annealing of microwires allows the manipulation of the magnetoelastic anisotropy of glass-coated microwires, and therefore enhances the DW velocity by extending the field range for single DW propagation, as well as enhancing DW velocity at a given magnetic field due to internal stress relaxation. The correlation between the annealing influence on the local nucleation field distribution and a change of magnetic field dependence of DW velocity, in amorphous Fe-based microwires with two different compositions and geometric parameters has been observed. Consequently, the nature of local defects limiting the single DW propagation regime and damping the DW might be related with the internal stress distribution.

Under certain experimental conditions it is possible to manipulate the DW dynamics in a magnetic wire in a field-driven regime and observe controllable DW collisions. This controllable DW collision can be realized in different parts of the wire. Such DW collisions can be used to release pinned domain walls with weak external fields. The control of DW dynamics is essential for advanced race track memory devices.

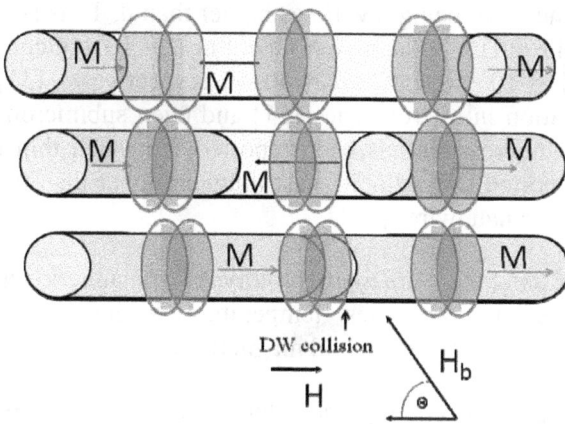

**Figure 66.** Scheme of the propagation of two DWs at three different times (t1<t2<t3) from the ends of a microwire sample in opposite directions, resulting in a DW collision in the middle of the sample. Reprinted with permission from [142]. Zhukov, J. M. Blanco, A. Chizhik, M. Ipatov, V. Rodionova, and V. Zhukova, Manipulation of domain wall dynamics in amorphous microwires through domain wall collision, *J. Appl. Phys.*, 114, 043910, (2013). Copyright [2013], AIP Publishing LLC (Fig. 5).

## 5.5. Applications of Microwires with Magnetic Bistability

### 5.5.1. Magnetic Sensors and Devices Based on Magnetic Bistability and Domain Wall Propagation

Magnetic domain wall propagation becomes a hot topic of research because of possibility of applications in magnetic devices, such as

Magnetic Random Access Memory, Integrated Circuits, Hard Disks, Domain-wall Logics, etc. [127-131]. High velocity of domain wall propagation observed in glass-coated microwires can be attractive to transmit the information along the magnetic microwire, like it was observed recently in wires of submicrometer diameter [131, 132, 136, 137]. Clear advantage of glass-coated microwires is that the domain wall velocity is few times higher, achieving few km/s [137]. The main problem is that generally the structure of propagating DW in magnetically soft microwires is far from abrupt and quite complex [131, 132]. As mentioned above, the characteristic width of the head to head DW, $\delta$, depends on few parameters, such as applied magnetic field, H, magnetic anisotropy constant, K, and microwires metallic nucleus diameter, d, being few times higher than d. Thus at H = 60 A/m $\delta \approx 65d$, while at H = 300A/m, $\delta \approx 40$ d. For K = $10^4$ erg/cm$^3$ $\delta/d \approx 13.5$, while for K = $10^3$ erg/cm$^3$ $\delta/d = 40$–50 respectively [131]. Recently DW propagation in micrometric [151] and even submicrometric [152] glass-coated microwires has been reported. For such thin microwires diameters reported DW velocity is still much higher than for the planar submicrometric nanowires [151].

On the other hand there are many prototypes of magnetic sensors based on magneto-clastic properties, temperature dependence of magnetic properties, magnetic softness, and magnetic surveillance.

Thus the magnetic bistability characterized by the sharp voltage peaks induced by abrupt change of the magnetic flux during the large Barkhausen jump is very convenient for the design of magnetic sensors.

Few of such applications have been mentioned and explained in previous reviews [8, 18]. Among them magnetoelastic sensor of the level of liquid designed using the very sensible stress dependence of the hysteresis loop and particularly of the coercivity in nearly-zero magnetostriction CoMnSiB microwires [8, 18], magnetic surveillance system consisting on magnetic tags containing bistable microwires with plurality of coercivities. As mentioned above, the glass-covered microwires technology allows tuning the internal stresses within the metallic nucleus opening a new way to engineer the switching field of these composite materials. Magnetic bistability with extended range of switching fields and high stress sensitivity of magnetic parameters (switching, field) give rise to various technological applications of tiny magnetic wires coated by glass [8, 18].

One of the applications is based on wide range of coercivity, which can be obtained in microwires owing to its strong dependence on geometric dimensions and heat treatment. It was realized in the method of magnetic codification using magnetic tags [8, 18]. The tag contains several microwires with well-defined coercivities, all of them characterized by a rectangular hysteresis loop. Once the magnetic tag submitted to the *ac* magnetic field, each particular microwire is remagnetized at different magnetic field giving rise to an electrical signal on a detecting system (see Fig. 67). The extended range of switching fields obtained in Fe-rich microwires gives a possibility to use a big number of combinations for magnetic codification.

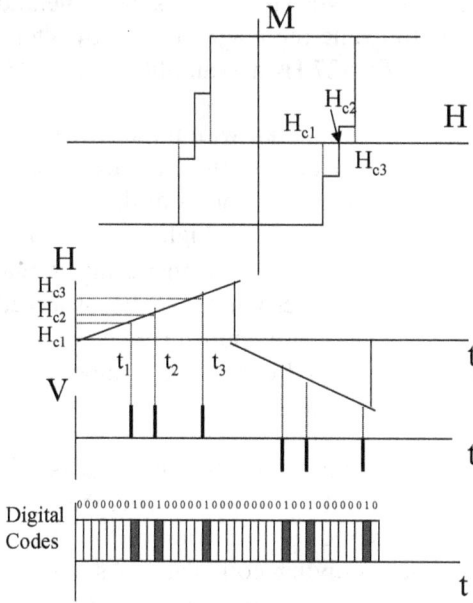

**Figure 67.** Schematic representation of the encoding system based on magnetic bistability of the microwires. Reprinted with permission from Ref. [8] V. Zhukova, M. Ipatov and A Zhukov, Thin Magnetically Soft Wires for Magnetic Microsensors, *Sensors*, 9, 9216-9240, 2009 doi:10.3390/s91109216 (Fig. 13).

The magnetic bistability looks quite attractive for EAS applications. On the reader other hand the distance between and the tag is the limiting factor for some applications. It was demonstrated that the electromagnetic response of the pieces of Fe-based microwires with diameter $D = 15 - 30$ μm and several centimeter length can be detected at a distance more than 20 cm from the wire [153].

The electromagnetic response of the magnetic tags consisting of a piece of microwire with different diameters in applied alternating magnetic field has been studied as shown in Fig. 68 employing two square coils with opposite connection as receivers. The receiving coils with a side of $a = 20$ cm were made of 20 turns of thin copper wire. The low noise band amplifier with amplification ~ 100 and the transmission band in the range of frequencies 500 Hz – 12 kHz has been used to amplify the electromotive force signals (EMF) generated in the receiving coils. The low frequency spectrum analyzer CF 5210 and digital oscilloscope were used to record the EMF signals. A small solenoid with a length 5 cm has been used to excite the magnetic tags by alternating magnetic field with a frequency $f = 327$ Hz and amplitude $H_0 = 5$ Oe.

In the absence of the magnetic tag within the solenoid the EMF signal contains the components at the exciting frequency and its second and third harmonics, as well as the harmonics of the power supply at 50 Hz. The noise voltage at a frequency higher than 1 kHz was around ~20 $\mu V/Hz^{1/2}$. However, when the magnetic tag is placed within the solenoid its electromagnetic response can be observed by means of digital oscilloscope as a series of short positive and negative impulses. The corresponding EMF signal contains a number of harmonics up to the frequencies higher than 10 kHz.

During the experiments performed the amplitudes of 5-th, 7-th and 11-th harmonics of the exciting frequency averaged over the 5 seconds time interval were measured as a function of the distance between the magnetic tag and the receiving coil plane. As Fig. 68 b shows, the magnetic tag can be moved within a working area along the guiding line perpendicular to the coil plane and fixed at a desirable angle with respect to the coil plane in any point between 0 (point near the receiving coils) and 50 cm far from the coil plane. Two configurations were studied in the experiment. In the first configuration the guiding line starts at the center of one of the receiving coil and the solenoid with the magnetic tag is oriented along the guiding line, perpendicular to the coil plane. In the second configuration the origin of the guiding line is located in the middle between the receiving coil centers, however the solenoid with magnetic tag is oriented horizontally and parallel to the receiving coil plane. Figs. 69 a, 69 b show the characteristic spectra of the EMF signals for the first tag configuration recorded at the distances Z equal 46 cm and 30 cm from the coil plane in the frequency range $f$ = 5 Hz – 5 kHz. It is found that for the distances Z $\leq$ 30 cm the main contributions to the measured signals were the harmonics generated by the magnetic tag during its magnetization reversal. However, the measured spectra contain also the contribution at the frequency 50 Hz and its odd harmonics up to 2 kHz.

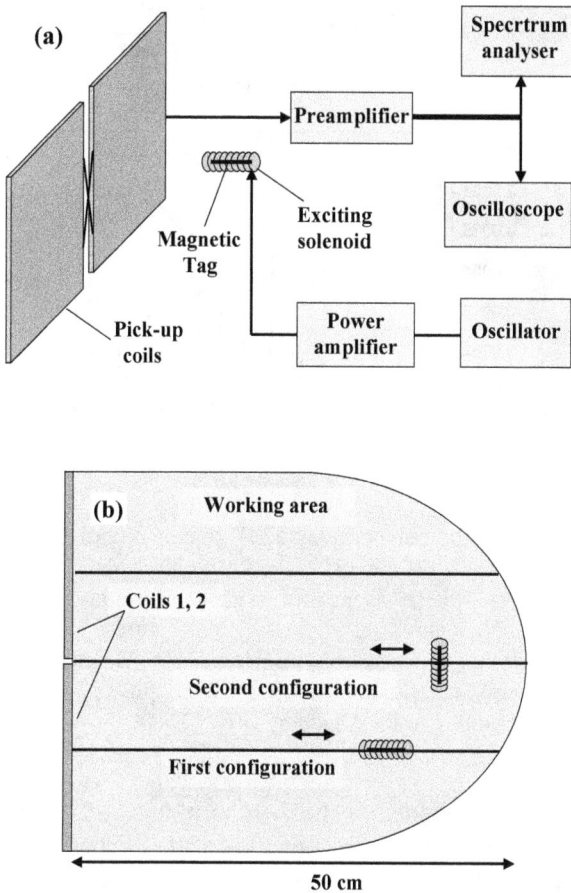

**Figure 68.** (a) Scheme of the installation to record the EMF signals of magnetic tags; (b) geometry of the working area near the receiving coils. Reprinted with permission from Ref. [154]. S. Gudoshnikov, N. Usov, A. Ignatov, V. Tarasov, A. Zhukov, and V. Zhukova, Ferromagnetic Microwire Usage for Magnetic Tags, *PIERS Proceedings*, Moscow, Russia, August 19-23, 2012, pp. 1274-1277. Copyright (2012) by The Electromagnetics Academy (Fig. 1).

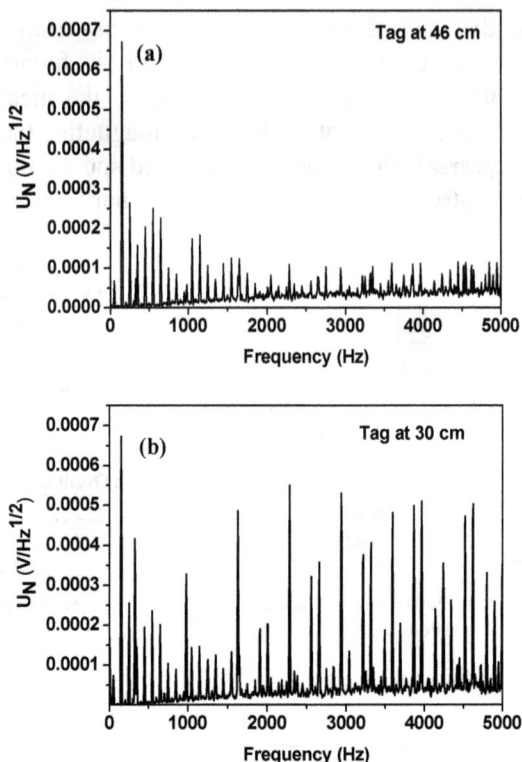

**Figure 69.** Spectra of the magnetic tag EMF signals recorded at the distances $z_1 = 46$ cm (a) and $z_2 = 30$ cm (b) far from the coil plane in the frequency range $f = 5$ Hz – 5 kHz. Reprinted with permission from Ref. [154]. S. Gudoshnikov, N. Usov, A. Ignatov, V. Tarasov, A. Zhukov, and V. Zhukova, Ferromagnetic Microwire Usage for Magnetic Tags, *PIERS Proceedings*, Moscow, Russia, August 19-23, 2012, pp. 1274-1277. Copyright (2012) by The Electromagnetics Academy (Fig. 2).

In Fig. 70 a the dependence of the 7-th harmonics of the magnetic tag EMF signal ($f_7 = 2289$ Hz) as a function of the distance from the coil plane is presented. One can see that even at the distance $z = 46$ cm the signal level is given by $\sim 90$ $\mu$V, that is 4 times higher than the noise level at this frequency. Similar data were obtained also for 11-th harmonics of the EMF signal at $f = 3597$ Hz. The results obtained were identical also for the case when the guiding line starts at the center of the second receiving coil. On the other hand, the signal level for the first tag configuration decreases considerably if the guiding line is shifted from the coil center to the point located in the middle between the receiving coil centers.

For the second configuration the dependence of the amplitude of the 7-th harmonics on the tag distance from the coil plane is shown in Fig. 70 a by the closed triangles. For this case the magnetic tag response can be confidently detected up to 50 cm from the coil plane.

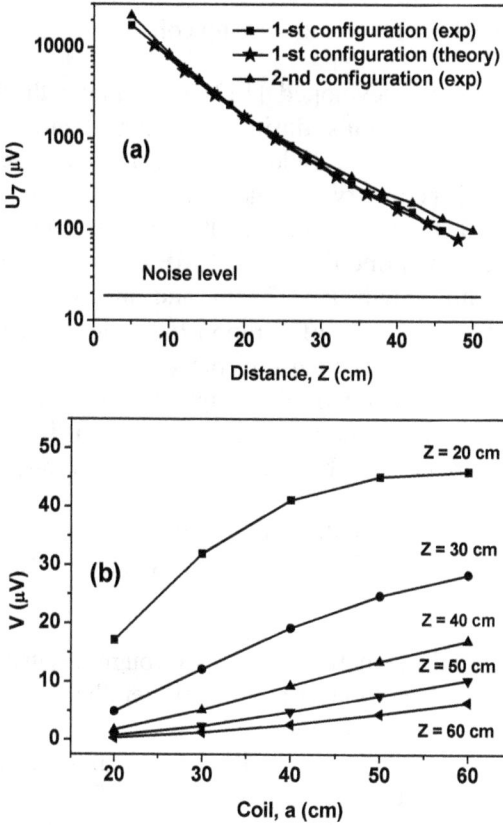

**Figure 70.** (a) The measured amplitude of 7-th harmonics of the magnetic tag EMF signals as a function of the distance of the tag from the receiving coil plane for the cases when the tag is oriented perpendicular (squares) or parallel (triangles) to the coil plane. The stars show the result of the theoretical calculation for the first tag configuration. (b) Theoretical calculation of the amplitude of 7-th harmonics as a function of the receiving coil size $a$ for various distances $Z$ of the magnetic tag from the receiving coil plane. Reprinted with permission from Ref. [154]. S. Gudoshnikov, N. Usov, A. Ignatov, V. Tarasov, A. Zhukov, and V. Zhukova, Ferromagnetic Microwire Usage for Magnetic Tags, *PIERS Proceedings*, Moscow, Russia, August 19-23, 2012, pp. 1274-1277. Copyright (2012) by The Electromagnetics Academy (Fig. 3).

One can see therefore, that for the given size of the receiving coils the magnetic tag having ~ 100 $\mu$m magnetic core diameter can be detected at the distances higher than 45 cm, irrespective of the magnetic tag orientation with respect to the receiving coil plane. However, at certain tag positions and orientations the EMF signal decreases considerably. One way to increase the sensitivity of the installation developed is to enlarge the sizes of the receiving pick-up coils.

The theoretical model developed [153] to estimate the EMF signals generated in receiving coils during tag magnetization reversals in external alternating magnetic field was used to study the influence of the receiving coil geometry on the amplitude of the EMF signals detected. Fig. 70 b shows the results of the calculations of the amplitude of the 7-th harmonics of the EMF coil's signal performed for magnetic tag with length l = 3 cm and magnetic core diameter d = 100 $\mu$m in the first tag configuration for various distances Z of the magnetic tag from the receiving coil plane. One can see in Fig. 70 b that for confident detection of the magnetic tag EMF signals at the distances up to 50 cm for the receiving coil plane it is desirable to use the receiving square coils with sizes a = 30 - 40 cm [154].

Consequently, glass-coated microwires with aforementioned dimensions might be used for the TAGs with detection distance of about 50 cm.

Creation of artificial magnetic structures through devitrification of the metallic nucleus and creation of additional metallic layers and forming arrays of microwires is also reported for creation of the tags [155]. Particularly, stray fields from the surrounding microwires can significantly change the magnetic response of wires arrays [7, 8, 155-157]. Such interaction takes place when few microwires are placed close to each other resulting in magnetostatic coupling among them. In a number of applications, arrays of closely spaced magnetic wires are of interest. In this case, their magnetic response will be affected by the stray magnetic fields from surrounding microwires [7, 8].

In most of previous paper the interaction of identical microwires has been studied [7, 8]. Recently we demonstrated that introduction of microwires with different character of hysteresis loop into the array can produce some novel effects and interesting peculiarities in magnetic response of the microwires array [155-157].

Interaction of identical Fe-rich microwires is illustrated in Fig. 71. As can be deduced from Fig. 71, single $Fe_{74}B_{13}Si_{11}C_2$ microwire exhibits change of the overall shape of hysteresis loops when $H_0$ exceeds some critical value (Fig. 71 a): at $H_0 > 330$ A/m we observed deviation from perfectly rectangular hysteresis loop typical for the magnetically bistable Fe-based microwire. This change of hysteresis loop shape at $H_0 > 330$ A/m must be related with the counterbalance between the sweeping rate, dH/dt and the switching time related with the time of domain wall propagation through the entire wire [155]. It is worth mentioning that general increase of coercivity with magnetic field frequency and amplitude has been previously observed elsewhere [155-159].

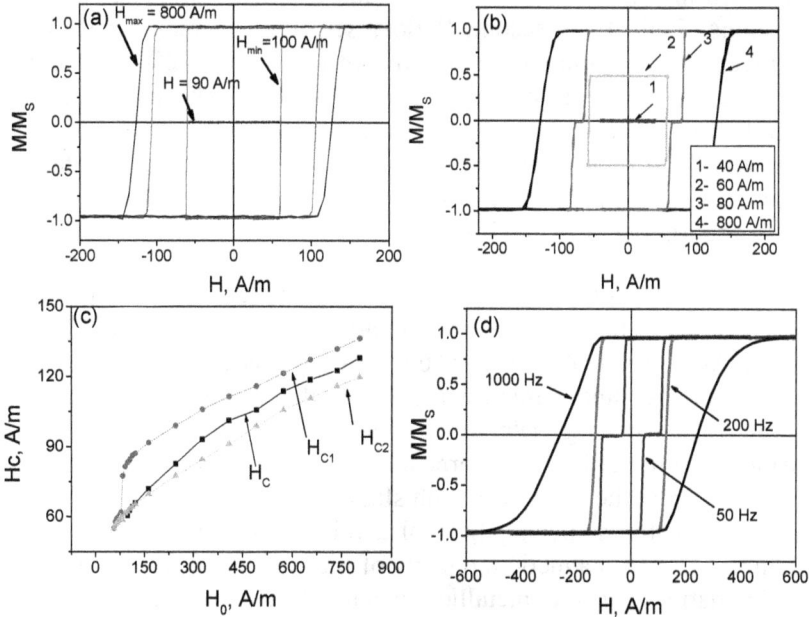

**Figure 71.** Hysteresis loops measured at different magnetic field amplitudes $H_0$ for single glass-coated $Fe_{74}B_{13}Si_{11}C_2$ microwire (a), array with two microwires (b); dependence coercivity $H_C$ on $H_0$ for single (solid line) and for array containing 2 microwires $H_{C1}$ and $H_{C2}$ (dot-line) (c); hysteresis loops of the two microwires measured at different magnetic field frequencies $f$ (d). Reprinted with permission from Ref. [156]. V. Rodionova, M. Ipatov, M. Ilyn, V. Zhukova, N. Perov, J. Gonzalez and A. Zhukov, Design of magnetic properties of arrays of magnetostatically coupled glass-covered magnetic microwires, *Phys. Status Solidi A*, 207, No. 8, 1954–1959 (2010). Copyright (2010) WILEY (Fig. 1).

Rectangular hysteresis loops typical for the magnetically bistable microwire were observed at magnetic field amplitude ranging from 100 A/m to 330 A/m. Below certain "critical" value of approximately $H_0$ = 100 A/m, the hysteresis loop disappears. Hysteresis loop, corresponding to $H_0$ = 800 A/m, the "critical" $H_0$ = 100 A/m (below this value the hysteresis loops disappears) and the $H_0$ = 90 A/m (at this value the hysteresis loop disappear, magnetization cannot be completed) are indicated in Figure 70 a by arrows.

As shown in Fig 71b, addition of second Fe-rich identical microwire results in splitting of the overall hysteresis loop and appearance of additional Barkhausen jump at 80 A/m$<H_0<$250 A/m, similarly to that previously observed [155-157]. As-compared with single microwire, the switching field of the first Barkhausen jump decreases, when the switching field of the second Barkhausen jump increases. Increasing magnetic field amplitude (approximately at $H_0 >$ 250 A/m), this splitting of hysteresis loop disappears (Fig.69b). The magnetic field amplitudes, $H_0$, at which the hysteresis loop slope disappears, are different for single microwire and for microwires array containing 2 samples. This dependence can be related with the stray field created by neighboring microwire [7, 8, 155-157].

Hysteresis loop splitting manifested as appearance of two steps in the hysteresis loops has been observed in array containing 2 microwires till $H_0$=80 A/m. At 80 A/m $<H_0<$ 60 A/m we observed hysteresis loop typical for the single microwire. Reduced value of switching field (below the switching field of each individual microwire) should be attributed to the effect of superposition of external magnetic field and stray field from the microwire with slightly higher switching field. This difference in the switching field of 2 microwires can be related with slightly different magnetic moments of the constituent microwires due to fluctuations of the metallic nucleus diameter, internal stresses, stresses arising during cutting the samples etc. Consequently switching fields of each microwire sample taken separately can be different. After switching of one of the microwires, superposition of the stray field of already re-magnetized microwire and of the external magnetic field is insufficient to switch the magnetization of the second wire with the larger moment. It results in the shift of the hysteresis loop along the M-axis for $M_{S1} \approx M_{S2}$ value, where $M_{S1}$ and $M_{S2}$ are saturation magnetization of the wires.

At elevated magnetic field amplitudes (490 A/m$<H_0<$ 800 A/m) this splitting of hysteresis loop disappears because the slope of the vertical regions of hysteresis loops is changed similar to the case of single microwire [158, 159]. This change of the slope of the vertical regions of hysteresis loops should be again attributed to the counterbalance between the sweeping rate, dH/dt = $4fH_o$ (i.e. increasing of $H_0$ results in faster change of magnetic field dH for the same time interval, dt) and the switching time related with the time needed for the domain wall propagation along the whole wire.

The magnetic stray fields of the neighboring microwires result in changing of the dynamic coercivities, $H_C$, of the array. The dependence of $H_C$ on magnetic amplitude field $H_0$ is shown in Fig. 70c for single microwire and two coupled microwires. The coercivities $H_{C1}$ and $H_{C2}$ in case of magnetization of the two microwires were determined in the following way:

$$H_{C1}=(H_{CI}+H_{CIII})/2, \; H_{C2}=(H_{CII}+H_{CIV})/2, \qquad (19)$$

where $H_{CI}$ - $H_{CIV}$ are the values of the magnetic fields taken at half-height of the hysteresis loops in the corresponding quarters I-IV of the coordinate space. Although the splitting of the hysteresis loop does not observed at elevated $H_0$, the difference in value of coercive force $H_C$ (in this case it is equivalent to the difference of hysteresis loops slope) in I and IV, II and III quarters, not distinguished by eye, indicate the presence of such interaction.

In fact the magnetization reversal is strongly affected by the magnetic field amplitude, $H_0$, and the frequency, $f$. The dependences of dynamic coercive force and overall hysteresis loop on $f$ and $H_0$ fit $H_C \sim \sqrt{fH_0}$ for the single Fe-based microwire [159]. Accordingly, the coercivity of the individual microwire follows this rule (see solid line in fig. 71c). For the array containing two $Fe_{74}B_{13}Si_{11}C_2$ microwires $H_C(f)$ dependence measured at fixed magnetic field amplitude ($H_0=800 \; A/m$) shows similar behavior (with the similar explanation) as for $H_C(H_o)$ dependence (Fig.71d.) Particularly, two-steps in the hysteresis loops are observed for $f<150$ Hz. At $f>150$ Hz this splitting of hysteresis loop disappears and one smooth-step has been observed for $150<f<1000$ Hz. Again, we assume that this change in character of hysteresis loop is related with the counterbalance of sweeping rate, dH/dt (increasing of $f$ results in faster change of magnetic field dH for the same time interval, dt) and the switching time related with the time necessary for the domain wall propagation through the whole wire. Consequently, both magnetic field amplitude, $H_0$, and frequency, $f$, affect overall shape of hysteresis loops.

Figs. 72 a and 72 b show the dependences of odd and even harmonics of the signal on magnetic field amplitude, respectively. It is worth mentioning that an abrupt increase in harmonics amplitudes is observed when the magnetic field amplitude is above a certain threshold depending upon $H_c$. The field plots of the odd harmonics having a "shelf" between fields of 60 and 90 A/m reflect the split hysteresis loops. Additionally, it should be noted that the amplitudes of even harmonics were substantially smaller than those for the odd ones. This reflects nearly symmetrical character of the magnetization process with respect to the field. The non-symmetry may be relatively important near the switching field as due to fluctuations the positive and negative switching fields are not exactly the same.

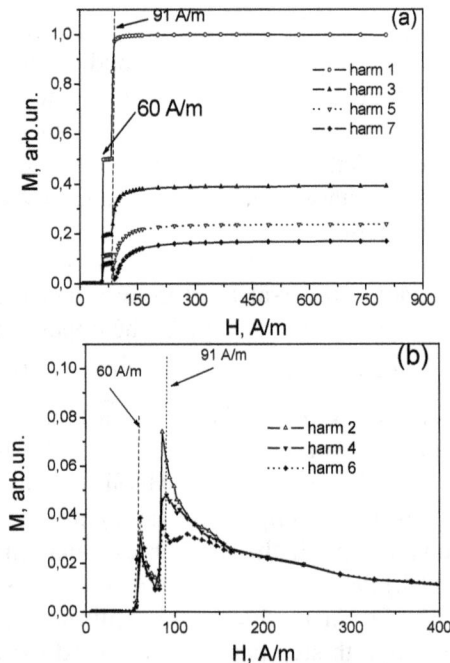

Figure 72. Dependences of odd harmonics (a) and even harmonics (b) on magnetic field amplitude in $Fe_{74}B_{13}Si_{11}C_2$ microwires. Reprinted with permission from Ref [157] V. Rodionova, M. Ipatov, M. Ilyn, V. Zhukova, N. Perov, L. Panina, J. Gonzalez and A. Zhukov, Magnetostatic interaction of glass-coated magnetic microwires, *J. Appl. Phys.*, 108 (2010) 016103. Copyright [2010], AIP Publishing LLC. (Figs. 1 b and 1 c).

Figure 73 a shows the hysteresis loop of single $Co_{67}Fe_{3.9}Ni_{1.5}B_{11.5}Si_{14.5}M_{0.6}$ microwire and the array consisting of 2 $Co_{67}Fe_{3.9}Ni_{1.5}B_{11.5}Si_{14.5}M_{0.6}$ microwires. It is seen that the presence of the second wire causes some increase in the effective anisotropy as the magnetization slope is decreased.

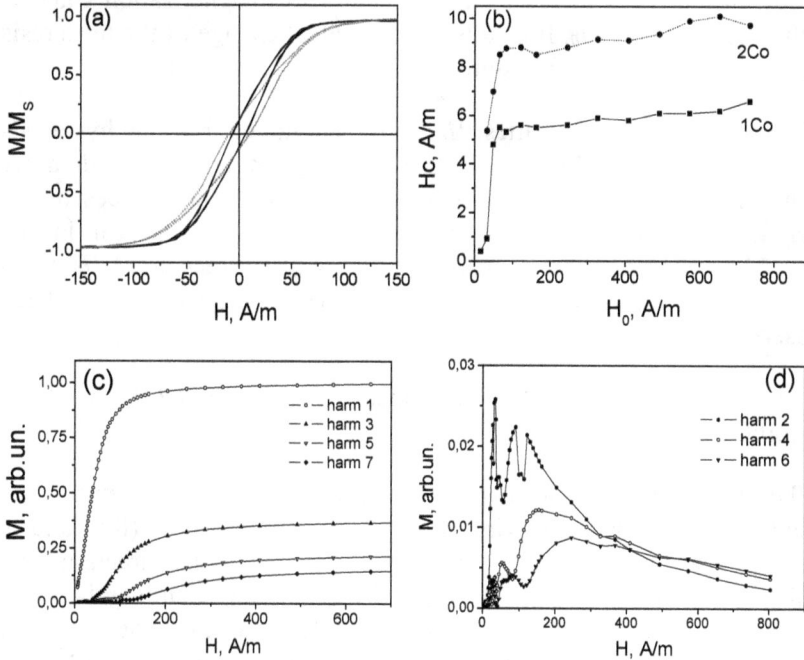

**Figure 73.** Hysteresis loops of the glass-coated $Co_{67}Fe_{3.9}Ni_{1.5}B_{11.5}Si_{14.5}M_{0.6}$ microwires (a) and dependence of coercivity $H_C$ of single (solid line) and two (dot-line) wires on magnetic field amplitude (b). Odd and even harmonics vs. field are given in (c) and (d), respectively. Reprinted with permission from Ref. [156]. V. Rodionova, M. Ipatov, M. Ilyn, V. Zhukova, N. Perov, J. Gonzalez and A. Zhukov, Design of magnetic properties of arrays of magnetostatically coupled glass-covered magnetic microwires, *Phys. Status Solidi A*, 207, No. 8, 1954–1959 (2010). Copyright (2010) WILEY (Fig.2 a and 2b) and from Ref [157] V. Rodionova, M. Ipatov, M. Ilyn, V. Zhukova, N. Perov, L. Panina, J. Gonzalez and A. Zhukov, Magnetostatic interaction of glass-coated magnetic microwires, *J. Appl. Phys.*, 108 (2010), 016103. Copyright [2010], AIP Publishing LLC. (Figs. 2 b and 2 c).

For these Co-based microwires with circular anisotropy, this changing can be explained by the increasing of effective demagnetization factor of the array as-compared with single microwire that is evidence of the existence of the magnetostatic interaction in the microwires array. The overall shape of the hysteresis loops does not depend on the number of microwires if the magnetic field amplitude, $H_0$, is above some critical value allowing magnetic saturation. This magnetic field amplitude, $H_0$, is 50 and 100 A/m for the single and two microwires, respectively. Further decreasing of $H_0$ results in changing of the shape of hysteresis loops attributed to the fact that the saturation magnetization can't be achieved for lower $H_0$. It results in the gradual change of the hysteresis loops shape: only minor hysteresis loops can be observed.

The values of the coercivities, $H_C$, were determined from the hysteresis loops for single $Co_{67}Fe_{3.9}Ni_{1.5}B_{11.5}Si_{14.5}M_{0.6}$ microwire and array consisting of two Co-based microwires and for all amplitudes of the magnetization field $H_0$. The dependence $Hc(H_0)$ is shown in the Fig. 71 b. Almost linear decreasing of $Hc$ with decreasing of $H_0$ down to 40 and 30 A/m (for one and two microwires, respectively) has been observed.

It should be mentioned that such linear dependence $Hc(H_0)$ has been theoretically predicted [159].

When the magnetic field amplitude, $H_0$, is below magnetic anisotropy field, $H_k$, of single microwire or the array of microwires (the field is insufficient to turn the magnetic moment off the circular direction), the value of coercivity abruptly drops. Owing to the coupling between microwires with circular anisotropy, this critical value of magnetic field amplitude $H_0$ is higher in the system consisted of two microwires than in single microwire.

Figs. 73 c and 73 d demonstrate the spectral characteristics of the magnetization process in non bistable microwires. As it could be expected from symmetry, the amplitude of even harmonics was substantially lower than those for odd ones. The field dependence of harmonics does not demonstrate any abrupt changes as in the case of magnetically bistable microwires. The magnitude of harmonics goes to saturation at fields much larger than the anisotropy fields reflecting a non-linear approach of the magnetization to saturation.

Fig. 74 a shows the hysteresis loops of the array consisting of one $Fe_{74}B_{13}Si_{11}C_2$ and one $Co_{67}Fe_{3.9}Ni_{1.5}B_{11.5}Si_{14.5}M_{0.6}$ microwires measured at different $H_0$. At $H_0 < 75$ A/m the loops are typical of those for Co-based microwires. For higher fields the hysteresis loops clearly reflect the superposition of magnetization processes including that by a large Barkhausen jump. Decreasing of $H_0$ below 120 A/m results in the change of the hysteresis loops shape. In a narrow range of $H_0$ below 120 A/m, there appears an interesting effect of a partial remagnetization of the bistable microwire. It is worth mentioning that remanent magnetization of Fe-based microwires is almost magnetic field independent, taking values, $-\mu_o M_r$ or $+ \mu_o M_r$. In contrary $\mu_o M_r$ of $Co_{67}Fe_{3.9}Ni_{1.5}B_{11.5}Si_{14.5}M_{0.6}$ microwire depends on magnetic field almost in a whole range of H (at least in low-field region). Therefore the presence of the Co-based microwire leads to the appearance of unusual hysteresis loop at low $H_o$ with the shape depending on $H_o$. We assume that at this magnetic field region the domain wall tears off the one end of the Fe-based microwire but does not have time to propagate through the whole wire because of the effect of stray field coming from Co-rich microwire. Further decreasing of $H_0$ below switching field of individual Fe-based sample leads to the disappearance of the influence Fe-based microwire and we observe the hysteresis loops typical for Co-based microwire.

The dependences of the coercivities $H_{CCo}$ and $H_{CFe}$ on magnetic field amplitude $H_0$ are presented in Figs. 71c and 73b. Both the $H_{CCo}$ and $H_{CFe}$ are larger than the coercivities of the single microwires because of the influence of stray field from neighboring microwire.

Considerable non-linearity of hysteresis loop can be appreciated (Fig. 74 a). Consequently, both odd and even harmonics can be enhanced through such magnetostatic interaction (Figs. 74 c, d). The spectral characteristics as functions of $H_0$ also reflect this combinational magnetization process. The odd harmonics gradually increase with the field until the switching field threshold is reached where they experience an abrupt jump (see Fig. 74 c). It is worth mentioning that in a narrow field range with $H_0 > 150$ A/m the magnitude of a higher harmonic may be larger than that of a lower one. The even harmonics (see Fig. 74d) show a relatively large increase near the switching field similar to that for a pair of bistable wires.

Finally, Fig. 75 shows hysteresis loop of the array containing three $Fe_{74}B_{13}Si_{11}C_2$ and one $Co_{67}Fe_{3.9}Ni_{1.5}B_{11.5}Si_{14.5}M_{0.6}$ microwires (set of 3 bistable and 1 non bistable microwires). The hysteresis loops of complex shapes reflecting the combinational magnetization process depending on the magnetization state in each wire are revealed. Generally, the description of the hysteresis of such array is similar to that containing just one $Fe_{74}B_{13}Si_{11}C_2$ and one $Co_{67}Fe_{3.9}Ni_{1.5}B_{11.5}Si_{14.5}M_{0.6}$ microwires. In this case it is essential that we observe 3 different regions: i) Typical for single Co-rich microwire, ii) Typical for one Fe-rich and one Co-rich microwire (see previous case) and iii) Region containing 3 Barkhausen jumps with influence of Co-rich microwire. It is worth mentioning that in this case the non-linearity of the hysteresis loop can be substantially enhanced.

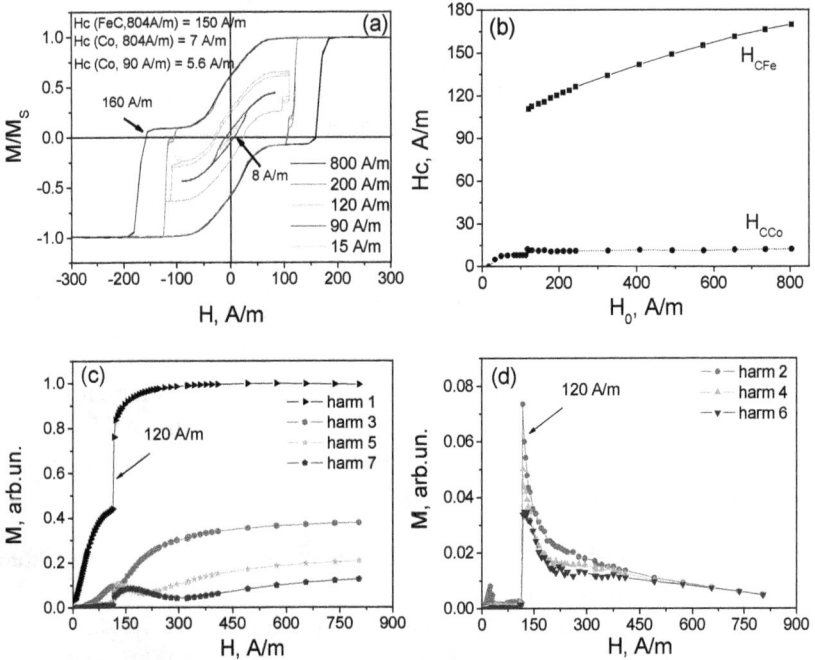

**Figure 74.** (a) Hysteresis loops of the $Fe_{74}B_{13}Si_{11}C_2$ + $Co_{67}Fe_{3.9}Ni_{1.5}B_{11.5}Si_{14.5}M_{0.6}$ array; (b) dependence of coercivity $H_C$ on magnetic field amplitude $H_0$ for Fe-based (solid) and Co-based (dots) microwires; (c) dependences of odd harmonics on magnetic field amplitude and (d) dependences of even harmonics on magnetic field amplitude. Reprinted with permission from Ref. [156]. V. Rodionova, M. Ipatov, M. Ilyn, V. Zhukova, N. Perov, J. Gonzalez and A. Zhukov, Design of magnetic properties of arrays of magnetostatically coupled glass-covered magnetic microwires, *Phys. Status Solidi A*, 207, No. 8, 1954–1959 (2010). Copyright (2010) WILEY (Fig. 3).

The harmonical spectra also reflect this multistage magnetization process with regions of gradual changes, abrupt jumps and shelves. The analysis of the hysteresis in arrays of different wires could be decomposed into three main processes:

- Gradual magnetization of a non bistable wire due to local wall motion and magnetization rotation;

- Large Barkhausen jump with a total remagnetization of a bistable wire;

- Gradual magnetization of a non bistable wire assisting a partial Barkhausen jump in a bistable wire.

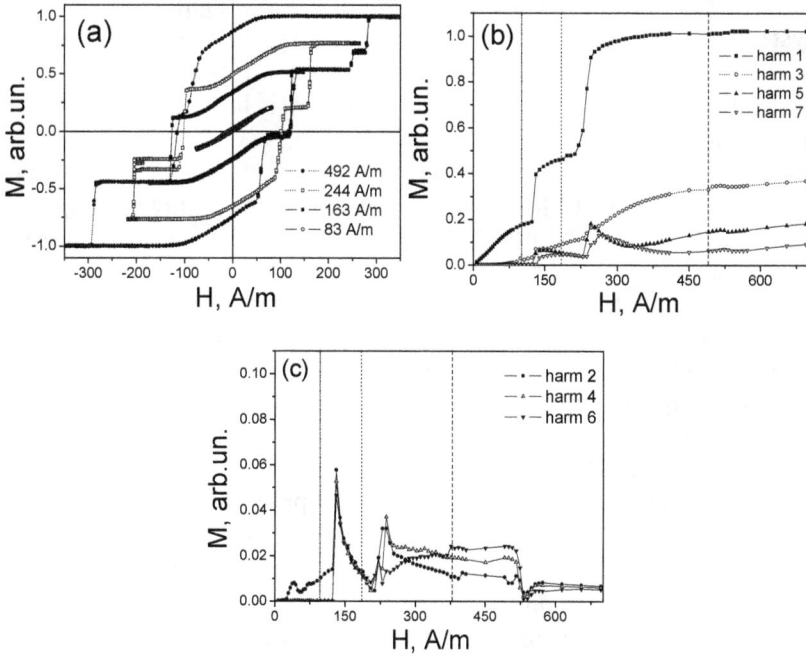

**Figure 75.** Hysteresis loops of 4-wire system consisting of three $Fe_{74}B_{13}Si_{11}C_2$ wires and a $Co_{67}Fe_{3.9}Ni_{1.5}B_{11.5}Si_{14.5}M_{0.6}$ wire in (a); odd and even harmonics vs. field are given in (b) and (c), respectively. Reprinted with permission from Ref. [157] V. Rodionova, M. Ipatov, M. Ilyn, V. Zhukova, N. Perov, L. Panina, J. Gonzalez and A. Zhukov, Magnetostatic interaction of glass-coated magnetic microwires, *J. Appl. Phys.*, 108 (2010), 016103. Copyright [2010], AIP Publishing LLC (Fig. 4).

It appears that utilizing arrays of bistable and non bistable magnetic wires allows us to tailor a highly complex magnetic response which can be decomposed into a number of characteristic processes. It is possible to discriminate those constituent stages not only from the hysteresis loop analysis but also from spectral analysis. Thus, in the 4-wire system, the 5-th and 7-th harmonics have characteristic sharp peaks where the magnetization changes abruptly and at those fields they become higher than the 3-d harmonic. It is worth mentioning that such non-linearity and enhanced harmonics are quite interesting for different security systems and magnetic labels [160].

Resuming, we showed that magnetic properties of the arrays consisted of Fe- and Co-rich microwires can be tailored through magnetostatic coupling among the microwires forming part of such arrays. Presence of neighbouring microwire (Fe- or Co-based) significantly affects hysteresis loop of the whole microwire array. In the case of mixed arrays containing Fe and Co-rich microwires we were able to obtain irregular hysteresis loops with unusual hysteresis loop shapes.

We have found that the interacting wires produce a highly complex magnetic response, in particular, when two types of microwires with bistable (Fe-based) and non bistable (Co-based) magnetization process are combined. In this case, the harmonic spectrum as a function of the field amplitude has a characteristic behaviour which can be easily identified by comparing the higher and lower harmonics. Together with conventional methods, such as thermal treatment, designing of arrays containing different types of microwires can be used for tailoring of their magnetic properties.

Resuming, we overviewed the magnetic properties of glass-coated microwires suitable for the magnetic sensor applications. Reduced dimensionality makes them quite interesting for the microsensors applications. At certain conditions (right composition, geometry, adequate annealing conditions) these microwires present high GMI effect. Magnetic field dependence of GMI ratio can be tailored either controlling magnetoelastic anisotropy of as-prepared microwires or by heat treatment or using magnetostatic interaction between microwires. Composite character of such microwires results in appearance of additional magnetoelastic anisotropy. Heat treatment is the efficient method of tailoring of magnetic properties and GMI effect of such microwires. Selection of proper chemical composition, geometry and adequate conditions of annealing allows achieving of high GMI effect. Studies of diagonal and off-diagonal MI tensor components of glass-coated microwires have shown the great potential of these materials for microminiaturized magnetic field sensor application.

Observed low-field GMI hysteresis can be suppressed by the bias electrical current. Low field GMI hysteresis has been observed and explained in terms of helical magnetic anisotropy of microwires.

Magnetostatic interaction within the linear array can significantly modify the magnetic response and the GMI effect of microwire arrays.

Magnetic microwires present a number of interesting effects, such as induction of the transversal anisotropy in Fe-rich microwires allowing observation of the stress-impedance effect and stress sensibility of overall hysteresis loop shape.

A single and large Barkhausen jump is observed for soft magnetic microwires with positive magnetostriction. The mechanism of the fast magnetization reversal is related with the remagnetization of inner single domain by fast domain wall propagation (DW velocity $v$ up to 2500 m/s). Below some critical magnetic field, $H_N$, determined by the microwires inhomogeneities, generally almost linear $v(H)$ dependence is found. This regime is controlled by the single DW propagation. When the applied magnetic field exceeds $H_N$, new reverse domains can be nucleated and consequently tandem remagnetization mechanism can be realized. The nucleation of new reversed domains is determined by natural or artificially created defects in the microwires. This results in significant decrease of the magnetization switching time and acceleration of magnetization switching in magnetically bistable microwires.

Manipulating of the magnetoelastic energy through application of tensile stress, changing the magnetostriction constant and internal stresses of studied microwires we significantly affects the domain wall dynamics in magnetically bistable microwires. In order to achieve higher DW propagation velocity at the same magnetic field and enhanced DW mobility special attention should be paid to decreasing of magnetoelastic energy. For low magnetostricitive samples DW velocity up to 2.5 km/s has been achieved.

These properties are quite attractive for the microsensors applications.

# Acknowledgements

Authors acknowledge collaboration with Dr. M. Ipatov, Dr. A. Chizhik, Dr. J.M. Blanco, Prof. J Gonzalez, Prof. M. Vázquez, Prof. L. Panina, Prof. S. Kaloshkin, Dr. M. Churiykanova, Dr. E. Varga, Dr. R. Zuberek, Dr. N. Usov, Dr. S. Gudoshnikov, Dr. C. Gracia, Dr. D. Mahnovsky, Dr. V. Rodionova, Mrs. K. Chichai, Mr. A. Talaat and other colleagues and collaborators.

This work was supported by the EU ERA-NET programme under project "SoMaMicSens" (MANUNET-2010-Basque-3), by EU under FP7 "EM-safety" project, by Spanish MICINN under Project MAT2010-18914, MAT2009-13108-C02-01 and by the Basque Government under Saiotek-2012 MEMFOMAG (S-PE12UN139).

# References

[1]. D. C. Giles, Recent advances and future directions in magnetic materials, *Acta. Mater.*, Vol. 51, 2003, pp. 5907–5939.

[2]. J. Durand, Magnetic Properties of Metallic Glasses in Topics in Applied Physics Volume 53, Glassy Metals II. Atomic Structure and Dynamics, Electronic Structure, Magnetic Properties, Editors: H. Beck and H.-J. Giintherodt, *Springer-Verlag*, Berlin Heidelberg New York Tokyo, 1983.

[3]. J. González and A. Zhukov, Amorphous magnetic materials for sensors, Encyclopedia of Sensors, Ed. by C. A. Grimes, E. C. Dickey, and M. V. Pishko, *American Scientific Publishers,* Vol. 1, 2006, pp. 79-103, ISBN: 1-58883-056-X.

[4]. G. Herzer, Amorphous and Nanocrystalline Materials, Encyclopedia of Materials: Science and Technology, ISBN: 0-08-0431526, *Elsevier Science Ltd*, 2001, pp. 149-157.

[5]. I. S. Miroshnichenko and I. V. Salli, A device for the crystallization of alloys at a high cooling rate, *Industrial Laboratory*, Vol. 25, 1959, pp. 1463-1466 in English, *Zav. Lab.,* Vol. 25, 1959, p. 1398.

[6]. P. Duwez, R. J. Williams, K. Klement, Continuous series of metastable solid solutions in Ag-Cu alloys, *Journal of Applied Physics*, Vol. 31, 1966, pp. 1136–1142.

[7]. H. Chiriac and T. A. Ovari, Amorphous glass-covered magnetic wires: preparation, properties, applications, *Progress in Material Science*, Vol. 40, 1997, pp. 333-407.

[8]. V. Zhukova, M. Ipatov and A Zhukov, Thin Magnetically Soft Wires for Magnetic Microsensors, *Sensors,* Vol. 9, 2009, pp. 9216-9240.

[9]. M. Vazquez, H. Chiriac, A. Zhukov, L. Panina and T. Uchiyama, On the state-of-the-art in magnetic microwires and expected trends for scientific and technological studies, *Physica Status Solidi A*, Vol. 208, 2011, pp. 493–501.

[10]. Magnetic Thin Films: Properties, Performance and Applications, Editor: John P. Volkerts, *Nova Science Publishers*, New York, ISBN: 978-1-61209-302-4, 2011, p. 409.

[11]. V. Zhukova, A. Zhukov, V. Kraposhin, A. Prokoshin and J. Gonzalez, Magnetic properties and GMI of soft melt-extracted magnetic

amorphous fibers, *Sensors and Actuators (A)*, Vol. 106, 2003, pp. 225-229.

[12]. Y. Honkura, Development of amorphous wire type MI sensors for automobile use, *Journal of Magnetism and Magnetic Materials*, Vol. 249, 2002, pp. 375-381.

[13]. K. Mohri, T. Uchiyama, L. P. Shen, C. M. Cai, L. V. Panina, Amorphous wire and CMOS IC-based sensitive micro-magnetic sensors (MI sensor and SI sensor) for intelligent measurements and controls, *Journal of Magnetism and Magnetic Materials*, Vol. 249, 2002, pp. 351-356.

[14]. L. V. Panina and K. Mohri, Magneto-impedance effect in amorphous wires, *Applied Physics Letters*, Vol. 65, 1994, pp. 1189-1191.

[15]. R. Beach and A. Berkowitz, Giant magnetic field dependent impedance of amorphous FeCoSiB wire, *Applied Physics Letters*, Vol. 64, 1994, pp. 3652-3654.

[16]. M.-H. Phan, H.-X. Peng, Giant magnetoimpedance materials: Fundamentals and applications, *Progress in Materials Science*, Vol. 53, 2008, pp. 323–420.

[17]. Y. Konno and K. Mohri, Magnetostriction measurements for amorphous wires, *IEEE Transactions on Magnetics*, Vol. 25, No. 5, 1989, pp. 3623-3625.

[18]. A. Zhukov, V. Zhukova, Magnetic properties and applications of ferromagnetic microwires with amorphous and nanocrystalline structure, *Nova Science Publishers*, New York, ISBN: 978-1-60741-770-5, 2009.

[19]. A. Zhukov, V. Zhukova, J. M. Blanco, A. F. Cobeño, M. Vazquez and J. Gonzalez, Magnetostriction in glass-coated magnetic microwires, *Journal of Magnetism and Magnetic Materials*, Vol. 258-259, 2003, pp. 151-157.

[20]. N. A. Usov, A. S. Antonov and A. N. Lagar`kov, Theory of giant magneto-impedance effect in amorphous wires with different types of magnetic anisotropy, *Journal of Magnetism and Magnetic Materials*, Vol. 185, 1998, pp. 159-173.

[21]. S. I. Sandacci, D. P. Makhnovskiy, L. V. Panina, K. Mohri, and Y. Honkura, Off-Diagonal Impedance in Amorphous Wires and Its Application to Linear Magnetic Sensors, *IEEE Transactions on Magnetics*, Vol. 35, 2004, pp. 3505-3510.

[22]. A. Zhukov, M. Ipatov, V. Zhukova, C. García, J. Gonzalez, and J. M. Blanco, Development of ultra-thin glass-coated amorphous microwires for HF magnetic sensor applications, *Physica Status Solidi (a)*, Vol. 205, No 6, 2008, pp. 1367-1372.

[23]. A. Zhukov, M. Ipatov, M. Churyukanova, S. Kaloshkin, V. Zhukova, Giant magnetoimpedance in thin amorphous wires: From manipulation of magnetic field dependence to industrial applications, *Journal of Alloys and Compounds,* Vol. 586, 2014, pp. S279–S286.

[24]. D. A. Allwood, G. Xiong, C. C. Faulkner, D. Atkinson, D. Petit and R. P. Cowburn, Magnetic Domain-Wall Logic, *Science,* Vol. 309, 2005, pp. 1688-1692.

[25]. M. Hayashi, L. Thomas, Ch. Rettner, R. Moriya, X. Jiang and S. Parkin, Dependence of Current and Field Driven Depinning of Domain Walls on Their Structure and Chirality in Permalloy Nanowires, *Physical Review Letters,* Vol. 97, 2006, 207205.

[26]. A. Zhukov, Domain wall propagation in a Fe-rich glass-coated amorphous microwire, *Applied Physics Letters,* Vol. 78, 2001, pp. 3106-3108.

[27]. N. L. Schryer and L. R. Walker, The motion of 180° domain walls in uniform dc magnetic fields, *Journal of Applied Physics,* Vol. 45, 1974, pp. 5406-5421.

[28]. R. Varga, A. Zhukov, V. Zhukova, J. M. Blanco and J. Gonzalez, Supersonic domain wall in magnetic microwires, *Physical Review B.,* Vol. 76, 2007, 132406.

[29]. A. P. Zhukov, M. Vazquez, J. Velazquez, H. Chiriac, V. Larin, The remagnetization process in thin and ultra-thin Fe-rich amorphous wires, *Journal of Magnetism and Magnetic Materials,* Vol. 151, 1995, pp. 132-138.

[30]. D. P. Makhnovskiy, L. V. Panina, and S. I. Sandacci, Tunable microwave composites based on ferromagnetic microwires, Book chapter in, Progress in Ferromagnetism Research, Editor: V. N. Murray, *Nova Science Publishers Inc.,* ISBN: 1-59454-335-6, USA, 2005.

[31]. H. X. Peng, F. X. Qin, M. H. Phan, Jie Tang, L. V. Panina, M. Ipatov, V. Zhukova, A. Zhukov, J. Gonzalez, Co-based magnetic microwire and field-tunable multifunctional macro-composites, *Journal of Non-crystalline solids,* Vol. 355, 2009, pp. 1380–1386.

[32]. V. Panina, M. Ipatov, V. Zhukova, A. Zhukov, and J. Gonzalez, Magnetic field effects in artificial dielectrics with arrays of magnetic wires at microwaves, *Journal of Applied Physics,* Vol. 109, 2011, 053901.

[33]. T. Ono, H. Miyajima, K. Shigeto, K. Mibu, N. Hosoito, T. Shinjo, Propagation of a Magnetic Domain Wall in a Submicrometer Magnetic Wire, *Science,* Vol. 284, 1999, pp. 468-470.

[34]. Y. Yoshizawa, S. Oguma and K. Yamauchi, New Fe-based soft magnetic alloys composed of ultrafine grain structure, *Journal of Applied Physics*, Vol. 64, 1988, pp. 6044-6046.

[35]. G. Herzer, Grain size dependence of coercivity and permeability in nanocrystalline ferromagnets, *IEEE Transactions on Magnetics*, Mag-26, 1990, pp. 1397-1402.

[36]. A. Hernando and M. Vázquez, Rapidly Solidified Alloys, Ed. H. H. Liebermann, *Marcel Dekker*, New York, 1993, p. 553.

[37]. J. Arcas, C. Gómez-Polo, A. Zhukov, M. Vázquez, V. Larin and A. Hernando, Magnetic properties of amorphous and devitrified FeSiBCuNb glass-coated microwires, *Nanostructured Materials*, Vol. 7, No. 8, 1996, pp. 823-834.

[38]. K. Mohri, F. B. Humphrey, K. Kawashima, K. Kimura and M. Muzutani, Large Barkhausen And Matteucci Effects in FeCoSiB, FeCrSiB, and FeNiSiB Amorphous Wires, *IEEE Transactions on Magnetics*, Mag-26, No. 5, 1990, pp. 1789-1793.

[39]. M. Vázquez, M. Knobel, M. L. Sánchez, R. Valenzuela and A. Zhukov, Giant magnetoimpedance effect in soft magnetic wires for sensor applications, *Sensors and Actuators A*, Vol. 59, 1997, pp. 20-29.

[40]. A. V. Ulitovski and N. M. Avernin, Method of fabrication of metallic microwire, *USSR Patent*, No. 161325, 19.03.64. Bulletin No. 7, p. 14.

[41]. A. V. Ulitovsky, I. M. Maianski, A. I. Avramenco, Method of continuous casting of glass coated microwire, *USSR Patent*, No. 128427, 15.05.60. Bulletin No. 10, p. 14.

[42]. E. Ya. Badinter, N. R. Berman, I. F. Drabenko, V. I. Zaborovsky, Z. I. Zelikovsky and V. G. Cheban, Cast micwories and its properties, *Shtinica*, Kishinev, 1973.

[43]. L. Kraus, J. Schneider and H. Wiesner, Ferromagnetic resonance in amorphous alloys prepared by rapid quenching from the melt, *Czech. J. Phys. B*, Vol. 26, 1976, pp. 601-602.

[44]. A. Zhukov, J. González, J. M. Blanco, M. J. Prieto, E. Pina and M. Vázquez, Induced Magnetic Anisotropy in Co-Mn-Si-B Amorphous Microwires, *Journal of Applied Physics*, Vol. 87, 2000, pp. 1402-1408.

[45]. A. S. Antonov, V. T. Borisov, O. V. Borisov, A. F. Prokoshin and N. A. Usov, Residual quenching stresses in glass-coated amorphous ferro- magnetic microwires, *J. Phys. D: Appl. Phys.*, Vol. 33, 2000, pp. 1161-1168.

[46]. H. Chiriac, T.-A Ovari and A. Zhukov, Magnetoelastic anisotropy of amorphous microwires, *Journal of Magnetism and Magnetic Materials*, Vol. 254-255, 2003, pp. 469-471.

[47]. J. Velázquez, M. Vazquez and A. Zhukov, Magnetoelastic anisotropy distribution in glass-coated microwires, *Journal of Material Research*, Vol. 11, No. 10, 1996, pp. 2499-2505.

[48]. M. Ilyn, V. Zhukova, C. Garcia, J. J. del Val, M. Ipatov, A. Granovskyand A. Zhukov, Kondo Effect and Magnetotransport Properties in Co-Cu Microwires, *IEEE Transactions on Magnetics*, Vol. 48, Issue 11, 2012, pp. 3532 – 3535.

[49]. A. Zhukov, J. Gonzalez and V. Zhukova, Magnetoresistance in thin wires with granular structure, *Journal of Magnetism and Magnetic Materials*, Vol. 294, 2005, pp. 165-173.

[50]. A. Zhukov, C. Garcia, M. Ilyn, R. Varga, J. J. del Val, A. Granovsky, V. Rodionova, M. Ipatov, V. Zhukova, Magnetic and transport properties of granular and Heusler-type glass-coated microwires, *Journal of Magnetism and Magnetic Materials*, Vol. 324, 2012, pp. 3558–3562.

[51]. R. Varga, T. Ryba, Z. Vargova, K. Saksl, V. Zhukova, A. Zhukov, Magnetic and structural properties of Ni-Mn-Ga Heusler-type microwires, *Scripta Materialia*, Vol. 65, Issue 8, 2011, pp. 703 – 706.

[52]. M. I. Ilyn, V. Zhukova, J. D. Santos, M. L. Sánchez, V. M. Prida, B. Hernando, V. Larin, J. González, A. M. Tishin, and A. Zhukov, Magnetocaloric effect in nanogranular glass coated microwires, *Physica Status Solidi (a)*, Vol. 205, No. 6, 2008, pp. 1378-1381.

[53]. V. Zhukova· A. M. Aliev, R. Varga, A. Aronin, G. Abrosimova, A. Kiselev and A. Zhukov, Magnetic properties and MCE in Heusler-type glass-coated microwires, *Journal of Superconductive and Novel Magnetism*, Vol. 26, Issue 4, 2013, pp. 1415-1419.

[54]. V. Zhukova, A. Chizhik, A. Zhukov, A. Torcunov, V. Larin and J. Gonzalez, Optimization of giant magneto-impedance in Co-rich amorphous microwires, *IEEE Transactions on Magnetics*, Vol. 38, Issue 5, part I, 2002, pp. 3090-3092.

[55]. K. R. Pirota, L. Kraus, H. Chiriac, M. Knobel, Magnetic properties and GMI in a CoFeSiB glass-covered microwire, *Journal of Magnetism and Magnetic Materials*, Vol. 221, 2000, L243-L247.

[56]. P. Aragoneses, A. Zhukov, J. Gonzalez, J. M. Blanco and L. Dominguez, Effect of AC driving current on Magneto-Impedance effect, *Sensors and Actuators* A, Vol. 81/1-3, 2000, pp. 86-90.

[57]. A. S. Antonov, N. A. Buznikov, M. M. Filatov, V. P. Goncharov, A. A. Rakhmanov, A. L. Rakhmanov, Effect of longitudinal AC magnetic field on frequency spectrum of voltage response of soft magnetic conductors, *Journal of Magnetism and Magnetic Materials*, Vol. 258–259, 2003, pp. 198–200.

[58]. A. F. Cobeño, A. Zhukov, J. M. Blanco and J. Gonzalez, Giant magneto-impedance effect in CoMnSiB amorphous microwires, *Journal of Magnetism and Magnetic Materials,* Vol. 234, 2001, L359-L365.

[59]. M. Knobel, M. Vazquez and L. Kraus, Giant magnetoimpedance, Handbook of Magnetic Materials (ed. K. H. Buschow), Chapter 5, Vol. 15., pp. 1-69, *Elsevier,* 2003.

[60]. D. Menard, M. Britel, P. Ciureanu, and A. Yelon, *Journal of Applied Physics,* Vol. 84, No5, 1998, pp. 2805-2814.

[61]. D. P. Makhnovskiy, L. V. Panina, and D. Mapps, Field-dependent surface impedance tensor in amorphous wires with two types of magnetic anisotropy: Helical and circumferential, *Physical Review B.,* Vol. 63, 2001, 144424-1—144424-17.

[62]. L. Landau and E. M. Lifshitz, Electrodynamics of Continuous Media, *Pergamon,* 1975.

[63]. H. Fujimori, K. I. Arai, H. Shirae, H. Saito, T. Masumoto and N. Tsuya, Magnetostriction of Fe-Co amorphous alloys, *Japanese Journal of Applied Physics,* Vol. 15, No. 4, 1976, pp. 705-706.

[64]. E. P. Harrison, G. L. Turney and H. Rowe, Electrical properties of wires of high permeability, *Nature,* Vol. 135, No. 3423, 1935, p. 961.

[65]. L. Kraus, M. Vázquez and A. Hernando, Creep-induced magnetic-anisotropy in a Co-rich amorphous wire, *Journal of Applied Physics,* Vol. 76, 1994, pp. 5343-5348.

[66]. K. Mohri, K. Kawashima, T. Kohzawa, Y. Yoshida and L. V. Panina, Magneto-inductance effect (MI effect) in amorphous wires, *IEEE Transactions on Magnetics,* Vol. 28, 1992, pp. 3150-3152.

[67]. J. Velázquez, M. Vázquez, D.-X. Chen and A. Hernando, Giant magnetoimpedance in non- magnetostrictive amorphous wires, *Physical Review B,* Vol. 50, 1994, pp. 16737- 16740.

[68]. A. P. Zhukov, The remagnetization process of bistable amorphous alloys, *Materials and Design,* No. 5, 1993, pp. 299-305.

[69]. K. Mohri, K. Kawashima, T. Kohzawa and H. Yoshida, Magneto-Inductive element, *IEEE Transactions on Magnetics,* Vol. 29, 1993, pp. 1245-1248.

[70]. K. L. Garcia and R. Valenzuela, Domain wall pinning, bulging and displacement in circumferential domains in CoFeBSi amorphous wires, *Journal of Applied Physics,* Vol. 87, 2000, pp. 5257-5259.

[71]. A. Yelon, D. Ménard, M. Britel and P. Ciureanu, Calculations of giant magnetoimpedance and ferromagnetic resonance response are

rigorously equivalent, *Applied Physics Letters,* Vol. 69, 1996, pp. 3084-3086.

[72]. A. S. Antonov, I. T. Iakubov, and A. N. Lagarkov, Nondiagonal impedance of amorphous wires with circular magnetic anisotropy, *Journal of Magnetism and Magnetic Materials,* Vol. 187, 1998, pp. 252–260.

[73]. D. Mahnovskiy, L. Panina. C. García, A. Zhukov and J. González, Experimental demonstration of tunable scattering spectra at microwave frequencies in composite media containing CoFeCrSiB glass-coated amorphous ferromagnetic microwires and comparison with theory, *Physical Review B*, Vol. 74, 2006, 064205-1-11.

[74]. W. S. Ament and G. T. Rado, Electromagnetic effects of spin wave resonance in ferromagnetic metals, *Physical Review*, Vol. 97, 1955, pp. 1558-1566.

[75]. V. A. Zhukova, A. B. Chizhik, J. Gonzalez, D. P. Makhnovskiy, L. V. Panina, D. J. Mapps and A. P. Zhukov, *Journal of Magnetism and Magnetic Materials,* Vol. 249, 2002, pp. 318-323.

[76]. M. Ipatov, V. Zhukova, J. M. Blanco, J. Gonzalez, and A. Zhukov, Off-diagonal magneto-impedance in amorphous microwires with diameter 6–10 μm and application to linear magnetic sensors, *Physica Status Solidi (a),* 205, No. 8, 2008, pp. 1779–1782.

[77]. A. Zhukov, M. Ipatov, C. García, J. Gonzalez, J. M. Blanco and V. Zhukova, Magnetic Properties and High–Frequency GMI Effect in Thin Glass-Coated Amorphous Wires, *AIP Conference Proceedings,* 1003, 2008, pp. 280-286.

[78]. A. Zhukov, M. Ipatov and V. Zhukova, Amorphous microwires with enhanced magnetic softness and GMI characteristics, *EPJ Web of Conferences*, Vol. 29, 2012, 00052.

[79]. M. Ipatov, V. Zhukova, A. Zhukov, J. Gonzalez, and A. Zvezdin, Low-field hysteresis in the magnetoimpedance of amorphous microwires, *Physical Review B,* Vol. 81, 2010, 134421.

[80]. C. García, A. Zhukov, V. Zhukova, M. Ipatov, J. M. Blanco and J. Gonzalez, Effect of Tensile Stresses on GMI of Co-rich Amorphous Microwires, *IEEE Transactions on Magnetics,* Vol. 41, 2005, pp. 3688-3690.

[81]. V. Zhukova, M. Ipatov, J. González, J. M. Blanco and A. P. Zhukov, Development of Thin Microwires With Enhanced Magnetic Softness and GMI, *IEEE Transactions on Magnetic,* Vol. 44, No. 11, Part 2, 2008, pp. 2958-3961.

[82]. V. Zhukova, A. F. Cobeño, A. Zhukov, J. M. Blanco, S. Puerta, J. González and M. Vázquez, Tailoring of magnetic properties of glass

coated microwires by current annealing, *Journal of Non-crystalline Solids*, 287, 2001, pp. 31-36.

[83].   A. Zhukov, Design of magnetic properties of Fe-rich glass – coated magnetic microwires for technical applications, *Advanced Functional Materials*, Vol. 16, Issue 5, 2006, pp. 675-680.

[84].   A. Zhukov, V. Zhukova, V. Larin, J. M. Blanco and J. Gonzalez, Tailoring of magnetic anisotropy of Fe-rich microwires by stress induced anisotropy, *Physica B,* 384, 2006, pp. 1-4.

[85].   V. Zhukova, V. S. Larin and A. Zhukov, Stress induced magnetic anisotropy and giant magnetoimpedance in Fe-rich glass/coated magnetic microwires, *Journal of Applied Physics,* Vol. 94, 2, 2003, pp. 1115-1118.

[86].   A. Zhukov, M. Churyukanova, L. Gonzalez, A. Talaat, V. Zhukova, B. Hernando, M. Ilyn, J. Gonzalez and S. Kaloshkin, Influence of Magnetoelastic Anisotropy on Properties of nanostructured Microwires, *Advanced Materials Research*, Vol. 646, 2013, pp. 59-66.

[87].   V. Zhukova, A. F. Cobeño, A. Zhukov, J. M. Blanco, V. Larin and J. González, Coercivity of glass-coated $Fe_{73.4-x}Cu_1Nb_{3.1}Si_{13.4+x}B_{9.1}$ ($0 \leq x \leq 1.6$) microwires, *Nanostructured Materials*, Vol. 11, 1999, pp. 1319-1327.

[88].   J. González, A. Zhukov, V. Zhukova, A. F. Cobeño, J. M. Blanco, A. R. de Arellano-López, S. López-Pombero, J. Martínez-Fernández, V. Larin and A. Torcunov, High coercivity of partially devitrified glass-coated *finemet* microwires: effect of geometry and thermal treatment, *IEEE Transactions on Magnetics,* Mag-36, 2000, pp. 3015-3017.

[89].   K. Mohri, T. Uchiyama and L. V. Panina, Recent advances of micro magnetic sensors and sensing application, *Sensors and Actuators,* A, Vol. 59, 2007, pp. 1–8.

[90].   T. Uchiyama, K. Mohri, and S. Nakayama, Measurement of Spontaneous Oscillatory Magnetic Field of Guinea-Pig Smooth Muscle Preparation Using Pico-Tesla Resolution Amorphous Wire Magneto-Impedance Sensor, *IEEE Transactions on Magnetics*, Vol. 47, No. 10, 2011, pp. 3070-3073.

[91].   K. Mohri, Y. Honkura, Amorphous Wire and CMOS IC Based Magneto-Impedance Sensors—Origin, Topics, and Future, *Sensor Letters,* Vol. 5, No. 2, 2007, pp. 267-270.

[92].   L. Ding, S. Saez, C. Dolabdjian, L. G. C. Melo, A. Yelon, and D. Ménard, Equivalent magnetic noise limit of low-cost GMI magnetometer, *IEEE Sensors Journal*, Vol. 9, No. 2, 2009, pp. 159-168.

[93]. L. Kraus, Theory of giant magneto-impedance in the planar conductor with uniaxial magnetic anisotropy, *Journal of Magnetism and Magnetic Materials,* Vol. 195, No. 3, 1999, pp. 764–778.

[94]. M. Ipatov, V. Zhukova, J. Gonzalez and A. Zhukov, Magnetoimpedance sensitive to DC bias current in amorphous microwires, *Applied Physics Letters,* Vol. 97, 2010, 252507.

[95]. D. Atkinson, P. T. Squire T, M. G. Maylin, J. Gore, An integrating magnetic sensor based on the giant magnetoimpedance effect, *Sensors and Actuators* A, Vol. 81, 2000, pp. 82–85.

[96]. S. Yabukami, H. Mawatari, N. Horikoshi, Y. Murayama, T. Ozawa, K. Ishiyama, K. I. Arai, A design of highly sensitive GMI sensor, *Journal of Magnetism and Magnetic Materials,* Vol. 290–291, 2005, pp. 1318–1321.

[97]. K. Nesteruk, M. Kuzminski, H. K. Lachowicz, Novel magnetic field meter based on giant magnetoimpedance (GMI) effect, *Sensors & Transducers,* Vol. 65, Issue 3, March 2006, pp. 515–520.

[98]. I. Giouroudi, H. Hauser, L. Musiejovsky, J. Steurer, Giant magnetoimpedance sensor integrated in an oscillator system. *Journal of Applied Physics,* Vol. 99, 2006, 08D906.

[99]. F. Alves, A. D. Bensalah, New 1D-2D magnetic sensors for applied electromagnetic engineering, *Journal of Materials Processing Technology,* Vol. 181, 2007, pp. 194–198.

[100]. H. Hauser, R. Steindl, C. Hausleitner, A. Pohl, J. Nicolics, Wirelessly interrogable magnetic field sensor utilizing giant magnetoimpedance effect and surface acoustic wave devices, *Transactions on Instrumentation and Measurement Technology,* Vol. 49, 2000, pp. 648–652.

[101]. R. Valenxuela, J. J. Freijo, A. Salcedo, M. Vazquez and A. Hernando, A miniature dc current sensor based on magnetoimpedance, *Journal of Applied Physics,* Vol. 81, 1997, pp. 4301–4303.

[102]. http://www.aichi-mi.com/

[103]. F. Qin, H.-X. Peng, Ferromagnetic Microwires Enabled Multifunctional Composite Materials, *Progress in Materials Science,* Vol. 58, 2013, pp. 183–259.

[104]. S. Gudoshnikov, N. Usov, A. Nozdrin, M. Ipatov, A. Zhukov and V. Zhukova (to be published in DICNMA conference proceedings).

[105]. L. Panina, M. Ipatov, V. Zhukova, J. Gonzalez and A. Zhukov, Tunable composites containing magnetic microwires, in Metal, ceramic and polymeric composites for various uses, Ed. by John Cuppoletti, Chapter 22, pp. 431-460, ISBN: 978-953-307-353-8

(ISBN 978-953-307-1098-3), *InTech - Open Access Publisher* (http://www.intechweb.org), Rijeka, Croatia.

[106]. J. B. Pendry, A. J. Holden, D J. Robbins, and W. J. Stewart, Low frequency plasmons in thin-wire structures, *Journal of Physycs: Condensed Matter*, Vol. 10, No. 22, 1998, pp. 4785-4809.

[107]. D. R. Smith, W. J. Padilla, D. C. Vier, S. C. Nemat-Nasser, and S. Schultz, Composite Medium with Simultaneously Negative Permeability and Permittivity, *Physical Review Letters*, Vol. 84, No. 18, May 2000, pp. 4184-4187.

[108]. A. N. Lagarkov & A. K. Sarychev, Electromagnetic properties of composites containing elongated conducting inclusions, *Physical Review B*, Vol. 53, No. 10, 1996, pp. 6318-6336.

[109]. D. P. Makhnovskiy and L. V. Panina, Field Dependent Permittivity of Composite Materials Containing Ferromagnetic Wires, *Journal of Applied Physics,* Vol. 93, No. 7, 2003, pp. 4120-4129.

[110]. V. Zhukova, M. Ipatov, A. Zhukov, R. Varga, A. Torcunov, J. Gonzalez and J. M. Blanco, Studies of magnetic properties of thin microwires with low Curie temperatures, *Journal of Magnetism and Magnetic Materials,* Vol. 300, 2006, pp. 16-23.

[111]. V. Zhukova, J. M. Blanco, M. Ipatov, A. Zhukov, C. Garcia, J. Gonzalez, R. Varga, A. Torcunov, Development of thin microwires with low Curie temperature for temperature sensors applications, *Sensors and Actuators* B, Vol. 126, 2007, pp. 318–323.

[112]. C. Heiden, H. Rogalla, Barkhausen jump field distribution of iron whiskers, *Journal of Magnetism and Magnetic Materials,* Vol. 26, 1982, pp. 275–277.

[113]. B. K. Ponomarev, A. P. Zhukov, Start field fluctuations of amorphous $Fe_5Co_{70}Si_{10}B_{15}$ alloys, *Fizika Tverdogo Tela*, Vol. 26, 10, 1984, pp. 2974-2979.

[114]. B. K. Ponomarev, A. P. Zhukov, Influence of temperature on the process of magnetization reversal of amorphous $Co_{70}Fe_5Si_{10}B_{15}$ alloys, *Acta Physica Polonica*, Vol. A28, No 2, 1985, pp. 259-263.

[115]. B. K. Ponomarev, A. P. Zhukov, Temperature effect on start field fluctuation distribution curve for amorphous $Co_{70}Fe_5Si_{10}B_{15}$ alloys, *Fizika Tverdogo Tela*, Vol. 27, No 2, 1985, pp. 444-448 .

[116]. D. X. Chen, C. Gomez-Polo and M. Vazquez, Magnetization profile determination in amorphous wires, *Journal of Magnetism and Magnetic Materials*, Vol. 124, 1993, pp. 262-268.

[117]. V. Zhukova, A. Zhukov, J. M. Blanco, J. Gonzalez and B. K. Ponomarev, Switching field fluctuations in a glass coated Fe-rich

amorphous microwire, *Journal of Magnetism and Magnetic Materials,* Vol. 249/1-2, 2002, pp. 131-135.

[118]. V. Zhukova, A. Zhukov, J. M. Blanco, J. Gonzalez, C. Gómez–Polo and M. Vázquez, Effect of applied stress on magnetization profile of Fe-Si-B amorphous wire, *Journal of Applied Physics* Vol. 93, 2003, pp. 7208-7210.

[119]. V. Zhukova, A. Zhukov, J. M. Blanco and J. Gonzalez, Effect of applied stress on remagnetization and magnetization profile of Co-Si-B amorphous wire, *Journal of Magnetism and Magnetic Materials,* Vol. 242-245, 2002, pp. 1439-1442.

[120]. V. Zhukova, A. Zhukov, J. M. Blanco, N. Usov and J. Gonzalez, Effect of applied stress on remagnetization and magnetization profile of Co-Si-B amorphous wire, *Journal of Magnetism and Magnetic Materials,* Vol. 258-259, 2003, pp. 189-191.

[121]. P. Gawronski, A. Zhukov, V. Zhukova, J. M. Blanco, J. González and K. Kulakowski, Distribution of switching field fluctuations in Fe-rich wires under tensile stress, *Applied Physics Letters,* Vol. 88, 2006, 152507.

[122]. R. Varga, A. Zhukov, M. Ipatov, J. M. Blanco, J. Gonzalez, V. Zhukova, P. Vojtaník, Thermal activation over complex energy barrier in bistable microwires, *Physical Review B,* Vol. 73, 2006, 054408-1-5.

[123]. D. A. Allwood, G. Xiong, C. C. Faulkner, D. Atkinson, D. Petit and R. P. Cowburn, Magnetic Domain-Wall Logic, *Science,* Vol. 309, 2005, pp. 1688-1692.

[124]. M. Hayashi, L. Thomas, Ch. Rettner, R. Moriya, X. Jiang, S. Parkin, Dependence of Current and Field Driven Depinning of Domain Walls on Their Structure and Chirality in Permalloy Nanowires, *Physical Review Letters,* Vol. 97, 2006, 207205(4).

[125]. D.-X. Chen, N. M. Dempsey, M. Vázquez and A. Hernando, Propagating domain-wall shape and dynamics in iron-rich amorphous wires, *IEEE Transactions on Magnetics,* 31, 1995, pp. 781-790.

[126]. W. Riehemann and E. Nembach, Tunneling of domain walls in ferromagnetic materials, *Journal of Applied Physics,* Vol. 55, 1984, pp. 1081-1091.

[127]. S. S. P. Parkin, Shiftable magnetic shift register and method of using the same, *U. S. Patent* No. US 683 400 5, 2004.

[128]. A. Kunza, S. C. Reiff, Enhancing domain wall speed in nanowires with transverse magnetic fields, *Journal of Applied Physics,* Vol. 103, 2008, 07D903.

[129]. A. Himeno, T. Ono, S. Nasu, T. Okuno, K. Mibu and T. Shinjo, *Journal of Magnetism and Magnetic Materials,* Vol. 272-276, 2004, 1577.

[130]. D. Atkinson, D. A. Allwood, G. Xiong, M. D. Cooke, C. C. Faulkner and R. P. Cowburn, Magnetic domain-wall dynamics in a submicrometre ferromagnetic structure, *Nature Materials*, Vol. 2, 2003, pp. 85-87.

[131]. S. A. Gudoshnikov, Yu. B. Grebenshchikov, B. Ya. Ljubimov, P. S. Palvanov, N. A. Usov, M. Ipatov, A. Zhukov, and J. Gonzalez, Ground state magnetization distribution and characteristic width of head to head domain wall in Fe-rich amorphous microwire, *Physica Status Solidi, A,* Vol. 206, No. 4, 2009, pp. 613–617.

[132]. P. A. Ekstrom and A. Zhukov, Spatial structure of the head-to-head propagating domain wall in glass-covered FeSiB microwire, *Journal of Physics D: Applied Physics*, Vol. 43, 2010, 205001.

[133]. Yu. Kabanov, A. Zhukov, V. Zhukova, J. Gonzalez, Magnetic domain structure of wires studied by using the magneto-optical indicator film method, *Applied Physics Letters*, Vol. 87, 2005, 142507.

[134]. K. J. Sixtus and L. Tonks, Propagation of large Barkhausen Discontinuities. II, *Physical Review*, Vol. 42, 1932, pp. 419-435.

[135]. M. Ipatov, N. A. Usov, A. Zhukov, J. Gonzalez, Local nucleation fields of Fe-rich microwires and their dependence on applied stresses, *Physica B,* 403, 2008, pp. 379–381.

[136]. M. Ipatov, V. Zhukova, A. K. Zvezdin and A. Zhukov, Mechanisms of the ultrafast magnetization switching in bistable amorphous microwires, *Journal of Applied Physics*, Vol. 106, 2009, 103902.

[137]. A. Zhukov, J. M. Blanco, M. Ipatov, A. Chizhik and V. Zhukova, Manipulation of domain wall dynamics in amorphous microwires through the magnetoelastic anisotropy, *Nanoscale Research Letters*, Vol. 7, 2012, 223.

[138]. K Richter, R Varga and A Zhukov, Influence of the magnetoelastic anisotropy on the domain wall dynamics in bistable amorphous wires, *Journal of Physics C: Condensed Matter*, Vol. 24, 2012, 296003.

[139]. A. Zhukov, J. M. Blanco, M. Ipatov, V. Rodionova, and V. Zhukova, Magnetoelastic Effects and Distribution of Defects in Micrometric Amorphous Wires, *IEEE Transactions on Magnetics*, Vol. 48, Issue 4, 2012, pp. 1324-1326.

[140]. J. M. Blanco, V. Zhukova, M. Ipatov, and A Zhukov, Effect of applied stresses on domain wall propagation in glass-coated amorphous microwires, *Physica Status Solidi, A,* Vol. 208, No. 3, 2011, pp. 545–548.

[141]. R. Varga, A. Zhukov, J. M. Blanco, M. Ipatov, V. Zhukova and J. Gonzalez, P. Vojtaník, Fast Magnetic Domain Wall in Magnetic Microwires, *Physical Review B*, Vol. 74, 2006, 212405-1-5.

[142]. R. Varga, K. Richter, A. Zhukov, V. Larin, Domain wall propagation in thin magnetic wires, *IEEE Transactions on Magnetics*, Vol. 44, No. 11, 2009, pp. 3925-3932.

[143]. L. V. Panina, M. Mizutani, K. Mohri, H. B. Humphrey and I. Ogasawara, Dynamics and Relaxation of Large Barkhausen Discontinuity in Amorphous Wires, *IEEE Transactions on Magnetics*, Vol. 27, No. 6, 1991, pp. 5331-5333.

[144]. L. V. Panina, M. Ipatov, V. Zhukova, A. Zhukov, Domain wall propagation in Fe-rich amorphous microwires, *Physica B*, Vol. 407, 2012, pp. 1442–1445.

[145]. G. S. D Beach, M. Tsoi, J. L. Erskine, Current-induced domain wall motion, *Journal of Magnetism and Magnetic Materials*, Vol. 320, 2008, pp. 1272–1281.

[146]. M. Ipatov, V. Zhukova, A. Zvezdin, J. Gonzalez, J. M. Blanco and A. Zhukov, Role of Defects on DomainWall Propagation in Magnetically Bistable Glass-Covered Microwires, *Journal of Superconductivy and Novel Magnetism*, Vol. 24, 2011, pp. 851–854.

[147]. K. Chichay, V. Zhukova, V. Rodionova, M. Ipatov, A. Talaat, J. M. Blanco, J. Gonzalez and A. Zhukov, Tailoring of domain wall dynamics in amorphous microwires by annealing, *Journal of Applied Physics*, Vol. 113, 2013, 17A318.

[148]. M. Ipatov, V. Zhukova, J. Gonzalez, and A. Zhukov, Annealing effect on local nucleation fields in bistable microwires, *Physica Status Solidi A*, Vol. 208, 2011, pp. 549-552.

[149]. V. Rodionova, V. Zhukova, M. Ilyn, M. Ipatov, N. Perov, A. Zhukov, The defects influence on domain wall propagation in bistable glass-coated microwires, *Physica B*, Vol. 407, 2012, pp. 1446-1449.

[150]. V. Zhukova, J. M. Blanco, M. Ipatov and A. Zhukov, Effect of transverse magnetic field on domain wall propagation inmagnetically bistable glass-coated amorphous microwires, *Journal of Applied Physics*, Vol. 106, 2009, 113914.

[151]. J. M. Blanco, V. Zhukova, M. Ipatov, and A. Zhukov, Magnetic Properties and Domain Wall Propagation in Micrometric Amorphous Microwires, *Sensor Letters*, Vol. 11, 2013, pp. 187-190.

[152]. H. Chiriac, S. Corodeanu, M. Lostun, G. Ababei, and T.-A. Óvári, Magnetic behavior of rapidly quenched submicron amorphous wires, *Journal of Applied Physics*, 107, 2010, 09A301.

[153]. S. Gudoshnikov, N. Usov, A. Zhukov, V. Zhukova, P. Palvanov and B. Ljubimov, O. Serebryakova, S. Gorbunov, Evaluation of use of magnetically bistable microwires for magnetic labels, *Physica Status Solidi A*, Vol. 208, No. 3, 2011, pp. 526-529.

[154]. S. Gudoshnikov, N. Usov, A. Ignatov, V. Tarasov, A. Zhukov, and V. Zhukova, Ferromagnetic Microwire Usage for Magnetic Tags, in *Proceedings of the PIERS,* Moscow, Russia, August 19-23, 2012, pp. 1274-1277.

[155]. V. Rodionova, M. Ipatov, M. Ilyn, V. Zhukova, N. Perov, J. Gonzalez and A. Zhukov, Tailoring of Magnetic Properties of Magnetostatically-Coupled Glass-Covered Magnetic Microwires, *Journal of Superconductivy and Novel Magnetism,* Vol. 24, No 1-2, 2011, pp. 541–547.

[156]. V. Rodionova, M. Ipatov, M. Ilyn, V. Zhukova, N. Perov, J. Gonzalez and A Zhukov, Design of magnetic properties of arrays of magnetostatically coupled glass-covered magnetic microwires, *Physica Status Solidi A*, Vol. 207, No. 8, 2010, pp. 1954–1959.

[157]. V. Rodionova, M. Ipatov, M. Ilyn, V. Zhukova, N. Perov, L. Panina, J. Gonzalez and A. Zhukov, Magnetostatic interaction of glass-coated magnetic microwires, *Journal of Applied Physics*, Vol. 108, 2010, 016103.

[158]. A. P. Zhukov, B. K. Ponomarev, Dependence of start field of amorphous Co- and Fe-rich alloys on frequency and magnetic field amplitude, *Fizika Tverdogo Tela*, Vol. 31, No 7, 1989, pp. 26-30.

[159]. A. Zhukov, M. Vázquez, J. Velázquez, C. Garcia, R. Valenzuela and B. Ponomarev, Frequency dependence of coercivity in rapidly quenched amorphous materials, *Journal of Material Science and Engineering A*, Vol. 226-228, 1997, pp. 753-756.

[160]. C. Moron, C. Aroca, M. C. Sanchez, A. Garcia, E. Lopez, P Sanchez, Application of flash annealed amirphous ribbons in security systems, *IEEE Transactions on Magnetics*, Vol. 31, 1995, pp. 906--909.

# Index

www.ingramcontent.com/pod-product-compliance
Lightning Source LLC
Chambersburg PA
CBHW050501190326
41458CB00005B/1389